Katharina Loewe

Arctic mixed-phase clouds

Macro- and microphysical insights with a numerical model

Wissenschaftliche Berichte des Instituts für Meteorologie und Klimaforschung des Karlsruher Instituts für Technologie (KIT)
Band 72

Herausgeber: Prof. Dr. Ch. Kottmeier

Institut für Meteorologie und Klimaforschung
am Karlsruher Institut für Technologie (KIT)
Kaiserstr. 12, 76128 Karlsruhe

Eine Übersicht aller bisher in dieser Schriftenreihe erschienenen Bände finden Sie am Ende des Buches.

Arctic mixed-phase clouds

Macro- and microphysical insights with a numerical model

by
Katharina Loewe

KIT Scientific Publishing

Dissertation, Karlsruher Institut für Technologie
KIT-Fakultät für Physik

Tag der mündlichen Prüfung: 26. Mai 2017
Referenten: Prof. Dr. sc. ETH Corinna Hoose
Prof. Dr. rer. nat. Thomas Leisner

Impressum

KIT Scientific Publishing

Karlsruher Institut für Technologie (KIT)
KIT Scientific Publishing
Straße am Forum 2
D-76131 Karlsruhe

KIT Scientific Publishing is a registered trademark
of Karlsruhe Institute of Technology.
Reprint using the book cover is not allowed.

www.ksp.kit.edu

Print on Demand 2017
Gedruckt auf 100 % Recyclingpapier mit dem Gütesiegel „Der Blaue Engel"

ISSN 0179-5619
ISBN 978-3-7315-0686-7
DOI 10.5445/KSP/1000070973

Arctic mixed-phase clouds

Macro- and microphysical insights with a numerical model

Zur Erlangung des akademischen Grades eines

DOKTORS DER NATURWISSENSCHAFTEN

von der Fakultät für

Physik

des Karlsruher Instituts für Technologie (KIT)

genehmigte

DISSERTATION

von

Dipl.-Met. Katharina Loewe (geb. Weixler)

aus Mühlacker

Tag der mündlichen Prüfung:	26. Mai 2017
Referent:	Prof. Dr. sc. ETH Corinna Hoose
Korreferent:	Prof. Dr. rer. nat. Thomas Leisner

Abstract

The Arctic is a very pristine and vulnerable region. In this region, climate change is dramatically apparent by larger temperature changes than in mid-latitudes. Furthermore, aerosol sources are rare in the Arctic resulting in low cloud condensation nuclei (CCN) concentrations. The Arctic surface is ice covered most of the time, but changes of the surface occur in the transition seasons spring and autumn. This leads to interactions with Arctic mixed-phase clouds. They are low-level clouds and regularly observed in these high latitudes. Moreover, they consist of supercooled droplets and ice particles at the same time. Depending on the surface type, Arctic mixed-phase clouds impact the surface radiative heating or cooling and the turbulent heating. Furthermore, they influence the boundary layer (BL) structure. Arctic mixed-phase clouds are influenced by changes in atmospheric conditions and are important for the surface energy balance. Processes that affect the life cycle of Arctic mixed-phase clouds are relatively poorly understood. Thus, field campaigns help to understand Arctic mixed-phase clouds and provide data for model studies to investigate microphysical and macrophysical processes related to the life cycle of Arctic mixed-phase clouds mechanistically.

This thesis provides new insights into macro- and microphysical properties of Arctic mixed-phase clouds: first, by comparing semi-idealized large eddy simulations with observations; second, by dissecting the influences of different surface types and BL structures on Arctic mixed-phase clouds; third, by elucidating the dissipation process of Arctic mixed-phase clouds; and finally by analyzing the main microphysical processes inside Arctic mixed-phase clouds.

The VERtical Distribution of Ice in Arctic clouds (VERDI) campaign was an aircraft campaign over the Beaufort Sea in spring 2012. Flights through clouds provide microphysical measurements inside mixed-phase clouds. Four different cases are investigated and compared to simulations with the COnsortium for Small-Scale MOdeling (COSMO) model. In general, the simulated cloud top, cloud base, and the liquid water content (LWC) within the cloud layer agree well with the observations. Differences are seen in the droplet size distribution. The simulated mean cloud droplet sizes are smaller than the observed droplet sizes due to the prescribed shape of the distribution in the two-moment scheme of cloud microphysics. The deviation of the observed and simulated droplet diameter profiles from the wet adiabatic droplet diameter profile is similar. In addition, the cloud top structure is analyzed. Observations and simulations show an inhomogeneous characteristic of the cloud top structure. The outcome of this study is a validated model providing the basis for systematic sensitivity studies, which are described in more detail below.

Different surface types cause variations in the sensible heat flux and in the latent heat flux. Thus, altered amount of moisture and heat can propagate into the BL. A first sensitivity study with three different surface types (sea ice, open lead, and sea water) shows that the structure of the BL and the surface type are important and influence the intensity of the surface fluxes. If the BL is well mixed to the surface, more moisture could reach the cloud layer and increase the cloud liquid water path. Particularly, the sea water surface influence the moisture transport by developing a well-mixed BL even when the BL is initially decoupled from the surface. This sensitivity study elucidates the influences of surface types and BL structures on Arctic mixed-phase clouds.

The Arctic Summer Cloud Ocean Study (ASCOS) campaign provides a large dataset of more than a month of observations. Measurements were done on board the Swedish icebreaker *Oden* as well as on a multi-year ice floe around 87°N, where *Oden* was moored for three weeks in summer 2008. A second sensitivity study focus on an episode at the end of the ice drift on 31 August 2008. A low-level mixed-phase cloud was quasi-persistent for around a week and dissipated in the evening of 31 August 2008. At the same time, observations showed a decrease in CCN concentration. Simulations are performed to identify the main contributors to dissipation of this Arctic mixed-phase cloud. A set of three different sensitivity experiments mimic large-scale advection of dry air and advection of aerosol. Increasing ice crystal number concentration (ICNC) lead to dissipation of the cloud only at an ICNC of $10\,\mathrm{l}^{-1}$, which is unrealistically high for the Arctic. Drying the atmospheric layers below the cloud base has no influence on the mixed-phase

cloud but drying the atmospheric layers above cloud top lead to a weakening of the liquid-dominated cloud layer. The main contributor for dissipation was found to be the reduction of cloud droplet number concentration (CDNC) to $2\,cm^{-3}$ in this specific observed Arctic mixed-phase cloud. While in reality, it is likely that a combination of different processes contribute to dissipation, this study provides a detailed analyis of how each process contributes to dissipation.
The same Arctic mixed-phase cloud during the ASCOS campaign as described before is analyzed to investigate the microphysical process rates inside clouds. Two groups of simulations, the diagnostic CDNC simulations with initial constant CDNC ($30\,cm^{-3}$ and $3\,cm^{-3}$) and the prognostic simulations with aerosol number concentrations of $30\,cm^{-3}$ and $80\,cm^{-3}$, prescribed CCN, and prognostic CDNC were performed. The ICNC is altered from $0\,l^{-1}$ to $1\,l^{-1}$. Different microphysical process rates are analyzed and the relevant processes are revealed. The warm phase process (autoconversion of cloud droplets to raindrops) is strongest near the cloud top, which is the region where the highest LWC is simulated. Additionally, the strongest sedimentation rate of rain is near the cloud top in all simulations, except for the simulation with a CDNC of $3\,cm^{-3}$, where a second layer at about 300 m above the surface and below the main cloud layer evolves. This cloud system finally dissipates. Varying the initial ICNC has no influence of the liquid droplet growth. Furthermore, the cloud lose condensate mainly via the liquid phase and not via the ice phase. The strongest ice phase process is the depositional growth of ice in the cloud layer, which indicates that the Wegener-Bergeron-Findeisen process has a strong influence on the simulated Arctic mixed-phase cloud. Near the surface, the ice particles sublimates. This precise sensitivity study evaluates microphysical processes inside an Arctic mixed-phase cloud and detect the most important processes.
On the whole, this thesis provides rigorous sensitivity analyses and elucidated important macrophysical and microphysical processes of Arctic mixed-phase clouds based on a validated high resolution model.

Contents

Glossary

Abbreviations

ASCOS	Arctic Summer Cloud Ocean Study
BACCHUS	Impact of Biogenic versus Anthropogenic emissions on Clouds and Climate: towards a Holistic UnderStanding
BL	boundary layer
CCN	cloud condensation nuclei
CCP	cloud combination probe
CDNC	cloud droplet number concentration
CE	condensation efficiency
COSMO	COnsortium for Small-Scale MOdeling
DWD	German Weather Service
ECMWF	European Centre for Medium-Range Weather Forecasts
ICNC	ice crystal number concentration
IN	ice nuclei
ISDAC	Indirect and Semi-Direct Aerosol Campaign
IWC	ice water content
IWP	ice water path
LES	large eddy simulations
LWC	liquid water content
LWP	cloud liquid water path
M-PACE	Mixed-Phase Arctic Cloud Experiment
MetUM	Met Office Unified Model
MMCR	Doppler millimeter cloud radar
MODIS	MODerate resolution Imaging Spectrometer

QNC	number density of cloud droplets
QV	specific humidity
RH	relative humidity
RWC	rain water content
RWP	rain water path
SCM	single-column model
SHEBA	Surface Heat Budget of the Arctic
SID3	Small Ice Detector mark 3
SWP	snow water path
TKE	turbulent kinetic energy
WVP	water vapor path
V0429	flight on 29 April 2012 during VERDI
V0514	flight on 14 May 2012 during VERDI
V0515	flight on 15 May 2012 during VERDI
V0517	flight on 17 May 2012 during VERDI
VERDI	VERtical Distribution of Ice in Arctic clouds
WBF	Wegener-Bergeron-Findeisen
WD	wind direction
WS	wind speed

Symbols

A	intercept of the generalized Γ-distribution
a_w	activity of water in aqueous solution
C_i	capacity of spherical particles
c_p	specific heat at constant pressure
D_V	diffusion of water vapour
$E_\mathrm{r,w}$	saturation vapour pressure over a curved water surface
$E_\mathrm{sat,w}$	saturation vapour pressure over a plane aqueous solution surface
$E_\mathrm{sat,s}$	saturation vapour pressure of the solution
$E_\mathrm{sat,w}$	saturation vapour pressure over a plane water surface
$f(x)$	number density size distribution function
F_V	ventilation coefficient
H_I	ice thickness
H_S	snow thickness
l	τ or LWP
κ_I	heat conductivity of ice
κ_S	heat conductivity of snow
K_h	turbulence coefficient of heat
K_m	turbulence coefficient of momentum
K_T	conductivity of heat
λ	slope parameter of the generalized Γ-distribution
L_iv	latent heat of sublimation
L_V	latent heat of condensation
m_S	mass of salt
m_W	mass of water
M_W	molar weight of water
M_S	molar weight of salt
μ	dispersion parameter of the generalized Γ-distribution
N	Brunt-Väisälä frequency
N_i	ice nucleation rate
N_Re	Reynolds number
N_Sc	Schmidt number
ν	number of ions into which a salt molecule dissociates in water
ν_S	shape parameter of the generalized Γ-distribution

ϕ_s	practical osmotic coefficient of salt in solution
p_0	standard reference pressure
P_l	spatial autocorrelation function of l
Q_I	heat flux through ice
Q_S	heat flux through snow
q_{tot}	total specific humidity (water vapour, water and ice)
$\mathscr{R}$	universal gas constant
r	radius of a spherical drop
$\mathscr{R}_d$	specific gas constant of dry air
r_{eff}	effective radius
ρ_l	one-dimensional inhomogeneity parameter of l
ρ_{LWP}	one-dimensional inhomogeneity parameter of LWP
ρ_s	density of the solution
ρ_τ	one-dimensional inhomogeneity parameter of τ
ρ_w	density of water
R_i	Richardson number
R_{i_c}	critical Richardson number
q_{tot_l}	total specific humidity (water vapour and water)
$\mathscr{R}_v$	specific gas constant of water vapour
S_i	saturation ratio over ice
σ_l	standard deviation of l
σ_{QNC}	standard deviation of QNC
$\sigma_{s/a}$	surface tension between the solution and humid air
S_l	one-dimensional inhomogeneity parameter of l
S_{LWP}	one-dimensional inhomogeneity parameter of LWP
$S_{v,w}$	saturation ratio
S_τ	one-dimensional inhomogeneity parameter of τ
$\sigma_{w/a}$	surface tension between water and humid air
T	temperature
τ	cloud optical thickness
T_f	freezing temperature of sea water
θ	potential temperature
θ_e	equivalent potential temperature
T_I	snow-ice interface temperature
T_S	snow-air interface temperature

w_e	entrainment velocity
ξ_l	decorrelation length of l
ξ_{LWP}	decorrelation length of LWP
z_{Tin}	initial temperature inversion height
z_i	inversion height

CHAPTER 1

Introduction

The Arctic is a vulnerable region and the Arctic system is extremely sensitive to climate changes (Curry et al., 1996). Furthermore, the Arctic atmosphere is cold and dry and the ocean is covered with first-year or multi-year sea ice (Stroeve et al., 2008). The sea ice inhibits the exchange of momentum and heat between the atmosphere and the relatively warm ocean, especially in winter. A sea ice covered ocean has an around 10 times higher surface albedo than the uncovered ocean. Therefore, the surface type has a strong influence on the surface energy budget. The Arctic climate is changing and the Arctic atmosphere warms faster than the global average and than in mid-latitudes (Serreze and Barry, 2011). Changes in moisture in the atmosphere, in clouds, and loss of sea ice intensify the Arctic warming (e.g. Manabe and Stouffer, 1980; Serreze et al., 2009; Serreze and Barry, 2011; Vihma et al., 2016).

A rapid sea ice decline in all season during the last decades, particularly in summer, changes the Arctic surface and impacts surface temperatures and moisture in the atmosphere (Stroeve et al., 2008; Vaughan et al., 2013; Serreze and Stroeve, 2015; Abe et al., 2016). Observations of the MODerate resolution Imaging Spectrometer (MODIS) show inter-annual lead occurrence in the Arctic (Willmes and Heinemann, 2016). Consequently, variations in atmospheric moisture has an effect on clouds in the Arctic atmosphere (Kay and Gettelman, 2009).

Clouds play an important role in the surface energy budget because they can reflect or absorb radiation from space. Due to their physical properties, they can cool or warm the surface and the atmosphere. Low-level clouds are common in the Arctic summer and autumn (Intrieri et al., 2002; Shupe et al., 2011). They can have a net

cooling effect in summer and a net warming effect on the surface around the year (Intrieri et al., 2002). Mixed-phase clouds are special, they are composed of ice crystals and supercooled liquid droplets. In this unstable mixture, the ice crystals can grow on the expense of the liquid droplets and the cloud can glaciate during a few hours. This process is called the Wegener-Bergeron-Findeisen (WBF) process (Wegener, 1911; Bergeron, 1935; Findeisen, 1938). Surprisingly, Arctic mixed-phase clouds can persist for several days (Morrison et al., 2012). This is due to several processes within the cloud and in the boundary layer (BL). The radiative cooling at cloud top is inducing turbulence in the cloud layer and this process helps to maintain the liquid layer of the cloud (e.g. Pinto, 1998). Furthermore, large scale moisture sources can be entrained at cloud top and prevents the liquid cloud layer from dissipation (Morrison et al., 2012). Additionally, the general synoptic scale forcing is usually weak in the Arctic and does hence not strongly influence the persistence of an Arctic mixed-phase cloud.
Field campaigns help to constrain and improve the simulation of Arctic mixed-phase clouds. Several campaigns were conducted in the past 20 years. Some of these are introduced in the following paragraph. One of the longest in time was the Surface Heat Budget of the Arctic (SHEBA) program with field experiments in the Arctic Ocean near Alaska from October 1997 to October 1998 (Uttal et al., 2002; Intrieri et al., 2002; Shupe et al., 2005). The goal was to gather data to improve our understanding of physical processes which are important for the surface energy budget and their interactions with air, sea, and ice. To look into Arctic mixed-phase clouds in more detail, the Mixed-Phase Arctic Cloud Experiment (M-PACE) was performed during the transition season in autumn 2004 on the North Slope of Alaska (Verlinde et al., 2007). Large eddy simulations (LES) were performed using data for instance from the Indirect and Semi-Direct Aerosol Campaign (ISDAC) at Barrow, Alaska in April 2008 (McFarquhar et al., 2011; Savre et al., 2014; Paukert and Hoose, 2014; Ovchinnikov et al., 2014). The focus was on interactions between cloud properties, such as cloud ice content or cloud water contents and BL structure in Arctic low-level mixed-phase clouds. The focus in this thesis is on observations from the VERtical Distribution of Ice in Arctic clouds (VERDI) campaign (Klingebiel et al., 2015) and the Arctic Summer Cloud Ocean Study (ASCOS) field campaign (Tjernström et al., 2014). The VERDI campaign was an aircraft campaign in April and May 2012 over the Beaufort Sea area north of Canada. Several flights through clouds provide an insight into profiles of

different cloud microphysical parameters and meteorological parameters, such as temperature and humidity. Simulations with focus on cloud formation, lifetime, and properties related to the BL structure of flights during the VERDI campaign are conducted. The ASCOS field campaign was done on board a ship and on an ice floe in summer 2008 around 87°N. During this field campaign a life cycle of an Arctic mixed-phase cloud was observed. These two campaigns are described in more detail in the corresponding chapters of this thesis (Chapter 4 and Chapter 6).
There are some challenges associated with simulating a mixed-phase cloud. A high resolution and a microphysics scheme are essential to represent the boundary layer and the gradients of various parameters, such as liquid water content, properly. One complex issue in modeling is the supercooled liquid layer near the cloud top, which is necessary for cloud top radiative cooling due to longwave emission (Chapter 2).
The COnsortium for Small-Scale MOdeling (COSMO) model is used with a high resolution in horizontal and vertical directions to investigate the micro- and macrophysical properties of Arctic mixed-phase clouds (Chapter 3). Furthermore, a closer look into the development and dissipation of an Arctic mixed-phase in the Arctic BL is given.
Moreover, this thesis analyzes simulations of a BL cloud above different surfaces like a closed sea ice surface or a more heterogeneous surface with an open lead surrounded by sea ice. It will investigate how different surfaces influence the microphysical properties of the Arctic mixed-phase cloud and the BL structure.
Finally, this thesis will provide insight into the dissipation of an Arctic mixed-phase cloud during the ASCOS field campaign. Additionally, the focus will be on the microphysical processes of the observed Arctic mixed-phase cloud during ASCOS.
In brief, the objectives of this thesis is to provide answers to the following key questions regarding Arctic mixed-phase clouds:

- How well does the model represent an Arctic mixed-phase cloud? How can different observations be used to evaluate the simulations? (Chapter 4)
- Do different surface types influence the cloud properties of an Arctic mixed-phase cloud? (Chapter 5)
- Why does an Arctic mixed-phase cloud dissipate? (Chapter 6)
- Which microphysical processes are relevant in Arctic mixed-phase clouds? (Chapter 7)

PART I

THEORY AND METHODS

CHAPTER 2

Arctic mixed-phase clouds

Mixed-phase clouds are typical in the Arctic and can occur during the whole year. They particularly appear in spring, late summer and autumn (e.g. Shupe et al., 2011). This chapter gives an overview about the Arctic environment, Arctic mixed-phase clouds and the conditions which are favourable for the persistence of an Arctic mixed-phase cloud. Furthermore, an insight into the microphysical processes in clouds is provided.

2.1 Arctic climate and boundary layer

The Arctic climate is influenced by cold air temperature, little moisture in the air, no or little solar radiation, and high surface albedo. All these conditions make the Arctic environment a unique place. Moreover, clouds play an important role in the Arctic system and influence the temperature, moisture and radiation. Arctic stratus clouds strongly affect the surface energy budget. They influence the surface energy balance by reflecting solar radiation and emitting longwave radiation. Most of the year the surface is covered by ice and is thus highly reflective. The surface reflectivity, the clouds, and the low position of the sun induce complex interactions of clouds and radiation (Curry et al., 1996). When the melt season starts in spring, more and more ocean surface appears and is contributing to the surface energy budget with input of heat and moisture. During this time, changes in the ice surface to open ponds or open leads change the radiative fluxes at the

surface. If more and more open water areas appear, reflection of solar radiation decreases. Therefore, radiative fluxes and heat fluxes can strongly vary from place to place and in time. They are most variable during the transition seasons spring and autumn. Hence, the effect of the surface fluxes on the Arctic climate varies throughout the whole year, too (Sedlar et al., 2011).
The Arctic boundary layer (BL) is mostly stable stratified and is often characterized by temperature inversions (e.g. Pinto, 1998). The ice covered ocean surface and weak solar radiation force these inversions. The strongest inversions occur during autumn and winter (Tjernström and Graversen, 2009). However, they occur throughout the whole year in the Arctic atmosphere. Therefore, the longwave emittance of an Arctic cloud is often at temperatures warmer than the surface below the cloud and Arctic clouds tend to warm the surface (Intrieri et al., 2002; Shupe and Intrieri, 2004). Of course, conditions and properties of Arctic clouds and the environment, for instance sun angle, cloud thickness or cloud height, are crucial to quantify the effect of clouds to the Arctic climate. The Arctic BL can either be coupled to or decoupled from the surface (Sotiropoulou et al., 2014) (Figure 2.1, equivalent potential temperature (θ_e) and total specific humidity (water vapour, water and ice) (q_{tot}) profiles). If radiative cooling at the cloud top generates a turbulent mixed layer, which reaches the surface and combines with the surface driven turbulence, the mixed-phase cloud is coupled to the surface. The surface driven turbulence depends on the surface type and its properties. For instance, temperature and roughness of the surface play a role in inducing turbulence. An ice surface is normally colder and rougher than a water surface. Also the transition zone from ice to a water surface may influence the surface turbulence production. During autumn, when the atmosphere cools, the surface starts to change, and more and more ice forms, the surface has a strong effect on the BL height (Schweiger et al., 2008; Sato et al., 2012). The more the sea ice decreases the more the BL increases. Besides that, observations show that decoupled BL clouds are generally higher than the coupled BL clouds, although no differences in cloud water and cloud thickness were observed (Sotiropoulou et al., 2014).

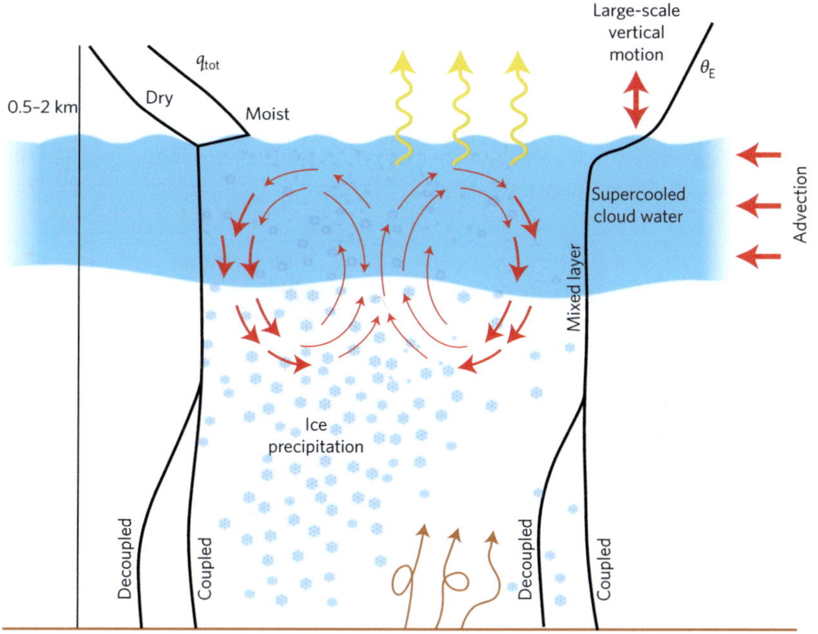

Figure 2.1: Scheme of processes of a persistent Arctic mixed-phase cloud. Characteristic profile of q_{tot} and the θ_e profiles are shown. Reprinted by permission from Macmillan Publishers Ltd: [Nature Geoscience] (Morrison et al., 2012), copyright (2012).

2.2 Characteristics of Arctic mixed-phase clouds

Aerosol sources are rare in the Arctic, especially in the high Arctic ($\geq 80\,^{\circ}$N). This unique environment affects mixed-phase clouds. Arctic mixed-phase clouds are low-level clouds with a cloud top at a height between 0.5 km and 2 km (Figure 2.1) (Morrison et al., 2012). The structure of an Arctic mixed-phase cloud typically comprises a liquid layer and an ice layer. Both layers together have approximately twice the thickness of the liquid layer. Between the liquid-dominated layer at cloud top and the layer containing only ice particles closer to the surface, a mixed-layer of ice crystals and supercooled cloud droplets develops. The liquid layer is around 300 m to 800 m deep (Shupe et al., 2008) and compared to mid-latitude clouds, the cloud droplet number concentration (CDNC) is lower. The Arctic CDNC is most of the time smaller than 100 cm^{-3} (Hobbs and Rangno, 1998). Typical temperatures of clouds in which cloud droplets and ice crystals coexist are -4 °C to -20 °C (Rangno and Hobbs, 2001).

The long persistence of these clouds is counter intuitive to the unstable mixture of supercooled water and ice crystals (Shupe et al., 2005). Normally, the ice particles would grow at the expense of the liquid droplets and the cloud would theoretically glaciate fast, known as the Wegener-Bergeron-Findeisen (WBF) process (Wegener, 1911; Bergeron, 1935; Findeisen, 1938). Instead, liquid droplets and ice particles can both grow at the expense of water vapour in the cloud at the same time (Korolev and Mazin, 2003; Korolev, 2007). This is possible, because the equilibrium vapour pressure over liquid is greater than the equilibrium vapour pressure over ice. Simultaneous growing of liquid droplets and ice particles may occur in updrafts, while simultaneous shrinking due to evaporation may occur in downdrafts and entrainment zones (Korolev and Mazin, 2003; Korolev, 2007). Thus, in convective clouds, the WBF process may be limited by the updraft velocity and may not be the only responsible mechanism in mixed-phase clouds.

Several processes in the Arctic environment help Arctic mixed-phase clouds to persist. These processes are weak synoptic scale forcing or large scale subsidence, radiative cooling and the induced turbulent mixing, and large scale moisture sources (Fig. 2.1) (Morrison et al., 2012; Shupe et al., 2013). Hence, Arctic mixed-phase clouds can persist for several days.

Among all these processes, the important small scale process in the cloud is radiative cooling. Droplets in the liquid cloud layer reflect the incoming solar radiation. The amount of reflected sunlight depends on the diameter of the droplets and the supply of liquid cloud water (Penner, 2004). Furthermore, the longwave cooling of the cloud air parcel near the cloud top leads to turbulent mixing and a cloud-driven mixed-layer develops (Solomon et al., 2011; Morrison et al., 2012; Shupe et al., 2013). Condensation can occur directly in the inversion layer due to radiative cooling (Solomon et al., 2011). Finally, the air parcels sink and the ice crystals inside the liquid layer can grow due to the expense of liquid water. The sedimentation of ice crystals starts in the mixed-layer of the cloud and forms a sink of the liquid water content (Morrison et al., 2012). In updrafts, cloud droplets

and ice crystals can grow, though the net liquid mass growth normally has the largest contribution (Shupe et al., 2008).
Large scale advection of moisture and humidity inversions at the cloud top may contribute weakly via entrainment as a source of moisture to the persistence of the cloud (Solomon et al., 2011; Sedlar et al., 2012; Tjernström et al., 2012; Nygård et al., 2014).

2.3 Microphysical processes in clouds

The atmosphere is composed of all three phases of water: the liquid, the solid, and the gas phase. Water vapour represents the gas phase, cloud droplets and raindrops the liquid phase, and ice particles the solid phase. The following section explains the warm and cold microphysical processes in clouds which lead to the growth of cloud droplets and ice particles.

2.3.1 Growth of cloud droplets

The growth process of cloud droplets is initiated by activation (Lamb and Verlinde, 2011). Before this, a first droplet has to form. Pure water droplets can theoretically form by cluster formation and condensation. This process, named homogeneous nucleation, needs a very high supersaturation in the atmosphere, which does not occur in nature (Wallace and Hobbs, 2006; Seinfeld and Pandis, 2016). Another way to build droplets is that water vapour condensates on aerosol particles, which then are totally or partly solved in water and form an initial droplet. The activation of cloud condensation nuclei (CCN) takes place at lower supersaturations and depends on the size of the aerosols and the composition of the solution. During the activation of a droplet, the droplet has to exceed a high energy barrier. Beyond this barrier named activation point, droplet growth is determined by the radius of curvature of the droplet. The activation point depends on the solute concentration of the droplet and the size of the droplet. The formerly solid particles which are dissolved in the water of the droplet decrease the saturation vapour pressure on the droplet surface and thus lead to a faster growing of the droplet compared to a pure water droplet. Depending on the chemical composition of the solution, droplets reach this point earlier or later while growing. The process of activation and further growth is described by the Köhler theory (Köhler, 1936). The theory combines the laws of Kelvin and Raoult. The Kelvin law (Equation (2.1)) for pure water describes the growth of droplets depending on the radius of curvature of the droplet. A curved surface has a higher saturation vapour pressure ($E_{\mathrm{r,w}}$) than a flat water surface ($E_{\mathrm{sat,w}}$). Therefore, droplets with a greater radius (r) grow by vapour diffusion on the expense of the smaller droplets (Pruppacher and Klett, 1997).

$$\frac{E_{\mathrm{r,w}}}{E_{\mathrm{sat,w}}} = \exp\left(\frac{2M_{\mathrm{w}}\sigma_{\mathrm{w/a}}}{\mathscr{R}T\rho_{\mathrm{w}}r}\right) \tag{2.1}$$

The surface tension and the compressibility of water are neglected in Equation (2.1). $\mathscr{R}$ is the universal gas constant ($= 8.31\,\mathrm{J\,mol^{-1}\,K^{-1}}$). ρ_{w} is the density of water and M_{w} the molar weight of water. The surface tension between water and humid air ($\sigma_{\mathrm{w/a}}$) amounts to $0.072\,\mathrm{J\,m^{-2}}$. In Equation (2.1) an equilibrium is assumed between the droplet and the environment. Thus, the Kelvin law can be expressed by the saturation ratio ($S_{\mathrm{v,w}}$) which describes the ratio between $E_{\mathrm{r,w}}$ and $E_{\mathrm{sat,w}}$:

$$\ln S_{\mathrm{v,w}} = \frac{2M_{\mathrm{w}}\sigma_{\mathrm{w/a}}}{\mathscr{R}T\rho_{\mathrm{w}}r} \tag{2.2}$$

The law of Raoult describes that the solution decreases the saturation vapour pressure over a plane aqueous solution surface ($E_{\mathrm{sat,w}}$) of the droplet and thus the activation of the droplets occur earlier (Pruppacher and Klett, 1997). The activity of water in aqueous solution (a_{w}) depends on the temperature (T), the saturation vapour pressure of the solution ($E_{\mathrm{sat,s}}$), and $E_{\mathrm{sat,w}}$. It is defined as follows:

$$a_{\mathrm{w}}(T) = \frac{E_{\mathrm{sat,s}}}{E_{\mathrm{sat,w}}} \tag{2.3}$$

Raoult‘s law (Equation (2.4)) takes the molar weight of salt (M_{s}), the mass of salt (m_{s}), the density of the solution (ρ_{s}) and number of ions into which a salt molecule dissociates in water (ν) into account. With the assumption of an ideal solution and with the practical osmotic coefficient of salt in solution (ϕ_{s}), Raoult‘s law is described as:

$$a_{\mathrm{w}} = \exp\left(-\nu\frac{m_{\mathrm{s}}M_{\mathrm{w}}}{M_{\mathrm{s}}(\frac{4\pi r^3}{3}\rho_{\mathrm{s}} - m_{\mathrm{s}})}\phi_{\mathrm{s}}\right) \tag{2.4}$$

The Köhler equation combines the size of the droplet and the water vapour saturation on the surface of the droplet (Equation (2.5)) (Pruppacher and Klett, 1997).

$$S_{\mathrm{v,w}} = a_{\mathrm{w}} K \tag{2.5}$$

K defines the Kelvin equation (Equation (2.2)). Under the presumption that the solution is sufficient diluted ($m_{\mathrm{s}} << $ mass of water (m_{w})) the $\sigma_{\mathrm{w/a}}$ is equal to surface tension between the solution and humid air ($\sigma_{\mathrm{s/a}}$). Furthermore, ρ_{s} is equal to ρ_{w} and $\phi_{\mathrm{s}} \approx 1$. The whole Köhler theory is described as following:

$$\ln S_{\mathrm{v,w}} = \frac{\mathrm{A}}{r} - \frac{\mathrm{B}}{r^3} \tag{2.6}$$

With:

$$\mathrm{A} = \frac{2 M_{\mathrm{w}} \sigma_{\mathrm{s/a}}}{\mathscr{R} T \rho_{\mathrm{w}} r}; \qquad \mathrm{B} = \nu \frac{m_{\mathrm{s}} M_{\mathrm{w}}}{M_{\mathrm{s}} (\frac{4\pi r^3}{3}) \rho_{\mathrm{w}}} \tag{2.7}$$

After the activation process, cloud droplets grow to raindrops by colliding with other cloud droplets (Figure 2.2). A larger droplet can collect a smaller one (accretion) or a faster droplet can catch a slower one (Lamb and Verlinde, 2011). Additionally, cloud droplets can grow by merging of droplets of different sizes. The latter process is called coalescence. During the collision process droplets can break up and build smaller droplets again. Collision and coalescence are the processes producing rain in warm clouds and the autoconversion represents the whole process of building raindrops out of cloud droplets. The loss of mass in clouds is due to sedimentation of raindrops. If the raindrops reach a certain size and mass they precipitate towards the surface.

2.3.2 Growth of ice particles

The temperature in clouds is often below the freezing temperature. Supercooled liquid droplets are still found in the atmosphere at temperatures below 0 °C and coexist with ice particles (Pruppacher and Klett, 1997).
Clean water droplets freeze at temperatures around -38 °C (Lamb and Verlinde, 2011). During freezing a critical cluster of ice molecules is built. This process is called homogeneous nucleation. However, homogeneous nucleation does not explain all the available ice in clouds. Ice particles exist at higher temperatures

up to 0 °C, too. Therefore, other processes come into play. Several ice-forming mechanisms like deposition nucleation, condensation, contact, and immersion freezing are describing various heterogeneous nucleation processes. In all these mechanisms, aerosols act as ice nuclei, interact with water vapour or liquid in the atmosphere, and favour the formation of ice particles (Lamb and Verlinde, 2011). Immersion freezing is assumed to be the major process in this thesis. It describes that a solid particle enters the liquid droplet before the freezing event occurs. Finally the droplet freezes and an ice particle is built.

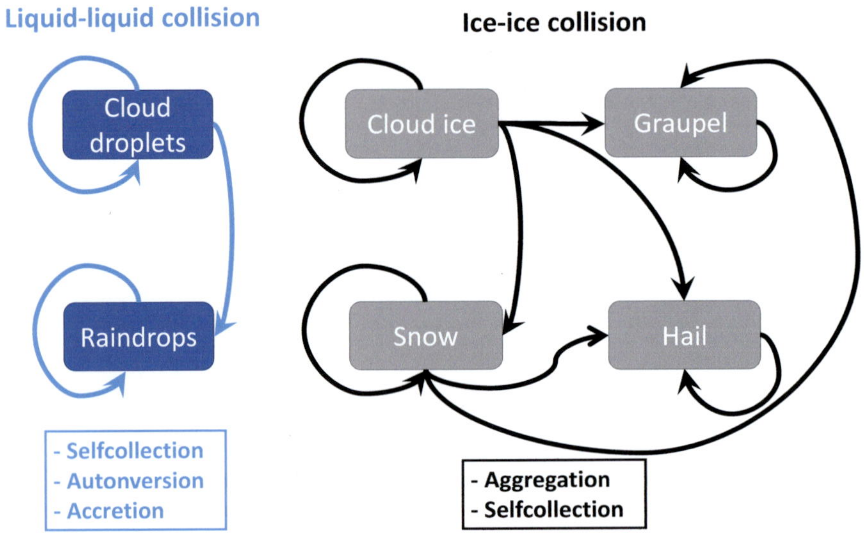

Figure 2.2: Overview of the collision processes of cloud droplets, raindrops, cloud ice, snow, graupel, and hail. The black lines inidicate ice-ice collision prozesses and the blue lines indicate liquid-liquid collision prozesses. Both processes are specified in the black and blue boxes, respectively.

Different processes in the atmosphere lead to the growth of ice particles (Seifert and Beheng, 2006). One possibility is that an ice particle collects other ice particles and then can grow further and build snowflakes with different sizes and shapes eventually (Figure 2.2). The aggregation of different ice particles to snowflakes is called autoconversion. Size, shape, and mass of the ice particles determine the velocity of sedimentation. For example, dendritical particles are slower than simple plate shaped particles of the same mass (Lamb and Verlinde, 2011).

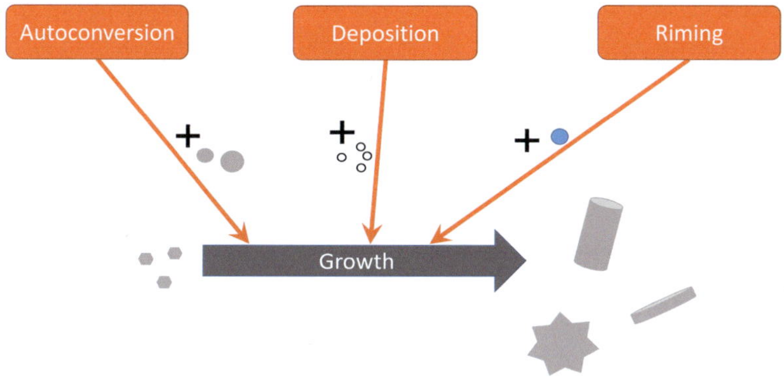

Figure 2.3: Overview of growth processes of ice particles. Autoconversion, deposition and riming are marked within the orange boxes. The grey colour marks ice, blue colour marks liquid and white indicates water vapour.

Further, ice particles can collide with supercooled liquid droplets, such that the liquid droplets usually instantaneously freeze at the ice particles. This is the riming process which occurs with cloud ice, snowflakes, graupel, and hail (Figure 2.3). The wet growth process is a special process which happens with hailstones. The fast collection of supercooled water by the hailstone lead to a warming of the hailstone due to latent heating. The hailstone grows by liquid water and the freezing process depends on the released latent heat. However, all these different types of ice hydrometeors can collide with supercooled liquid droplets and support the forming of particles with different sizes.
The depositional growth of an ice particle brings the water vapour in the atmosphere into play. The ice particles grow by diffusion of water vapour. The equilibrium vapour pressure over ice is lower than over liquid water. The growth of ice particles by vapour diffusion requires the supersaturation over ice and the supersaturation generated by adiabatic cooling at cloud top and updrafts limit the depositional growth (Lamb and Verlinde, 2011).
All these different processes contribute to the ice growth in a cloud and are summarized in Fig. 2.3.

CHAPTER 3

The COnsortium for Small-Scale MOdeling (COSMO) model

The COSMO model is a non-hydrostatic limited-area atmospheric prediction model, developed at the German Weather Service (DWD) for operational numerical weather prediction as well as for scientific purposes (Schättler et al., 2015). The model is based on primitive thermo-hydrodynamical equations which describe a compressible flow in a moist atmosphere. COSMO uses rotated geographical coordinates and generalized terrain following height coordinates. The grid structure is an Arakawa C-grid with a Lorenz vertical grid staggering. In the used setup, the time integration scheme is a third order Runge-Kutta scheme with time split treatment of acoustic and gravity waves. The radiation is calculated with a two-stream radiation scheme after Ritter and Geleyn (1992). Longwave and shortwave fluxes are calculated column wise. In the following sections, relevant components of the model, with importance to this thesis, are described.

3.1 Cloud microphysics

The two moment cloud microphysics scheme by Seifert and Beheng (2006) was developed for mixed-phase clouds and to improve the representation of clouds and precipitation in atmospheric models. Within this scheme the development of the number densities and the masses of six hydrometeors types, denoted cloud droplets, cloud ice, raindrops, snow, graupel, and hail, are predicted. The growth of droplets of clouds and rain are described by warm phase processes. Droplet activation, condensation, collection, and coalescence are parameterized and the drop size distribution composes of cloud droplets and raindrops (Seifert and Beheng, 2001). Raindrops have a radius greater or equal to a radius of $r^* = 40\,\mu$m. The cloud droplets are distinguished from the raindrops by the drop mass $x^* = 2.6 \times 10^{-10}$ kg.

The scheme is based on the partial power moments, M_c^k and M_r^k, of the number density size distribution function ($f(x)$) of droplets $f_\mathrm{w}(\mathrm{x})$. The first partial moment of $f_\mathrm{w}(\mathrm{x})$ is the number density of cloud droplets and raindrops, $M_\mathrm{c}^0 \equiv$ QNC and $M_\mathrm{r}^0 \equiv$ QNR, respectively (Seifert and Beheng, 2006). Furthermore, the second partial moment is the mass density of cloud droplets and raindrops, $M_\mathrm{c}^1 \equiv$ LWC and $M_\mathrm{r}^1 \equiv$ RWC.

$$M_\mathrm{c}^\mathrm{k} = \int_0^{\mathrm{x}^*} x^\mathrm{k} f_\mathrm{w}(x)\mathrm{d}x = \int_0^{\infty} x^\mathrm{k} f_\mathrm{c}(x)\mathrm{d}x, \tag{3.1}$$

$$M_\mathrm{r}^\mathrm{k} = \int_{\mathrm{x}^*}^{\infty} x^\mathrm{k} f_\mathrm{w}(x)\mathrm{d}x = \int_0^{\infty} x^\mathrm{k} f_\mathrm{r}(x)\mathrm{d}x \tag{3.2}$$

The whole size spectrum can be split in a number density size distribution function of cloud droplets $f_\mathrm{c}(x)$ and of raindrops $f_\mathrm{r}(x)$. $f_\mathrm{c}(x)$ and $f_\mathrm{r}(x)$ are described by generalized Γ-distributions,

$$f(x) = Ax^{\nu_\mathrm{s}} exp(-\lambda x^{\mu}) \tag{3.3}$$

with the two constant parameters ν_s and μ (Table 1 in Seifert and Beheng (2006)). They rely on the different hydrometeor types. A is the intercept, ν_s is the shape parameter, λ is the slope or scale parameter, μ is the dispersion parameter (Seifert and Beheng, 2006; Khain et al., 2015). A and λ can be described by the number and mass densities, for cloud droplets by the number density of cloud droplets (QNC) and by the liquid water content (LWC), respectively.

$$A = \frac{\mu \mathrm{QNC}}{\Gamma(\frac{\nu_\mathrm{s}+1}{\mu})} \lambda^{\frac{\nu_\mathrm{s}+1}{\mu}} \tag{3.4}$$

$$\lambda = \left[\frac{\Gamma(\frac{\nu_\mathrm{s}+1}{\mu})}{\Gamma(\frac{\nu_\mathrm{s}+2}{\mu})} \bar{x} \right]^{-\mu}, \bar{x} = \frac{\mathrm{LWC}}{\mathrm{QNC}} \tag{3.5}$$

Table 3.1: Coefficients of the relationships of diameter-mass and velocity-mass, which are used in the current model setup.

Particle	a	b	α	β
cloud droplets	0.124	1/3	3.75×10^5	2/3
raindrops	0.124	1/3	114.0137	0.234370
cloud ice	0.835	0.39	6.387	1/6
snowflakes	2.4	0.455	8.8	0.15
graupel	0.15	0.323	32.0	0.180
hail	0.129	1/3	56.0	0.21270

The different particles have different size and mass properties. Therefore, the relationships between diameter-mass and velocity-mass are parameterized as follows:

$$D(x) = ax^b \tag{3.6}$$

$$v(x) = \alpha x^{\beta} \left(\frac{\rho_0}{\rho}\right)^{\gamma} \tag{3.7}$$

The coefficients a, b, α, and β are constant and depend on the hydrometeor type (Seifert and Beheng (2006), Table 3.1). The effect of the air density (ρ) on the velocity of each hydrometeor is small. γ is 0.5 for large raindrops and for smaller hydrometeors γ is lower. This effect is considered by a linear fit of γ depending on the mean diameter of the hydrometeors (personal communication U. Blahak, DWD, Germany). ρ_0 is the surface air density.
The different ice phase hydrometeors are considered in the ice phase scheme. Parameterizations of the different ice growth processes are implemented (Figure 2.3). The depositional growth of ice particles is an important process in the simulations of Arctic mixed-phase clouds in this thesis, thus the model process will be described in more detail. The general growth rate of an individual ice particle ($\frac{dx_i}{dt}$) is given by

$$\left.\frac{dx_i}{dt}\right|_{dep} = \frac{4\pi C_i F_v(x_i) S_i}{\frac{\mathscr{R}_v T}{E_{sat,i}(T) D_v} + \frac{L_{iv}}{K_T T}\left(\frac{L_{iv}}{\mathscr{R}_v T} - 1\right)} \tag{3.8}$$

Equation 3.8 describes a vapour mass flux at the ice particle surface depending on the saturation ratio over ice (S_i) (Pruppacher and Klett, 1997; Seifert and Beheng, 2006). The capacity of spherical particles (C_i) is a function of the geometry of the ice particle ($C_i = \frac{D_i}{2}$) and the diffusion of water vapour (D_V) (3×10^5 m^2 s^{-1}), the specific gas constant of water vapour ($\mathscr{R}_v$) (461.51 J kg^{-1} K^{-1}), the latent heat of sublimation (L_{iv}) (2.834×10^6 J kg^{-1}), and the conductivity of heat (K_T) (2.5×10^{-2} J m K^{-1} s^{-1}) are constants. If an ice crystal has grown to a certain size it is important to consider the fall velocity by adding the ventilation coefficient (F_v). For spherical particles F_v is given by

$$F_v(x_i) = a_{v,i} + b_{v,i} N_{Sc}^{1/3} N_{Re}^{1/2} \tag{3.9}$$

with the ventilation coefficient $a_{v,i}$ and $b_{v,i}$, which depend on the diameter-mass relationship and velocity-mass relationship as well as on μ and ν of the generalized Γ-distribution ($f(x)$) (Equation (3.3)). F_v depends on the Schmidt number (N_{Sc}) (= 0.71 (Pruppacher and Klett, 1997)) and the Reynolds number (N_{Re}).

3.2 Sea ice model

Surface characteristics have an impact on the energy balance of the surface and sea ice has different radiative properties and roughness than for example an open water area. Therefore the COSMO model has a sea ice scheme which considers the thermodynamic processes in the ice layer and the possible snow layer on the ice layer (Figure 3.1) (Mironov and Ritter, 2004; Mironov, 2008; Schättler et al., 2015). The temperature profile is parameterized within these two layers and depends on the time and the vertical coordinate z (Eq. 3.10). For the parameterization the temperature at the snow-ice interface (T_I), the temperature at the snow-air interface (T_S), and (T_f), the freezing point of sea water, -1.7° C, is described as follows:

$$T(z,t) = \begin{cases} T_f - [T_f - T_I(t)]\phi_I(\zeta_I) & \text{at} - H_I(t) \le z \le 0 \\ T_I(t) - [T_I(t) - T_S(t)]\phi_s(\zeta_s) & \text{at} - [H_I(t) + H_s(t)] \le z \le -H_I(t) \end{cases} \tag{3.10}$$

H_I is the ice thickness and H_S the thickness of the snow. z is the vertical coordinate, being positive downward, and originates at the ice-water interface.
With the heat fluxes through snow (Q_S) and through ice (Q_I) (Eq. 3.11) two differential equations for the time dependent $T_I(t)$ and $H_I(t)$ build the system for the heat transfer. Introducing dimensionless functions $\phi_I \equiv (T_f - T)/(T_f - T_I)$

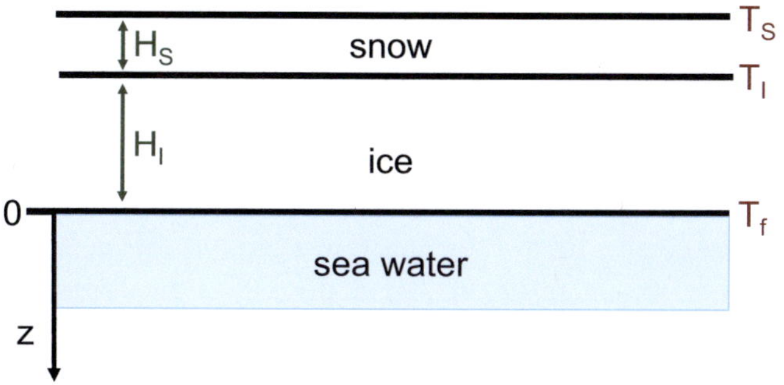

Figure 3.1: Scheme of the sea ice model configurations. The freezing point temperature (T_f) of sea water is defined at the sea water surface ($z = 0m$) and at the bottom of the ice. The ice layer with its thickness H_I is below the snow layer with a thicknes of H_S, with temperatures T_I and T_S, respectively. The vertical coordinate z is defined positive downward.

and $\phi_s \equiv (T_I - T)/(T_I - T_S)$, and dimensionless depths $\zeta_I \equiv -z/H_I$ and $\zeta_S \equiv -(z + H_I)/H_S$ suffice the boundary conditions $\phi_I(0) = 0$, $\phi_I(1) = 1$, $\phi_s(0) = 0$, and $\phi_s(1) = 1$.

$$\begin{aligned} Q_I &= -\kappa_I \frac{T_f - T_I}{H_I} \frac{d\phi_I}{d\zeta_I}, \\ Q_s &= -\kappa_s \frac{T_I - T_S}{H_S} \frac{d\phi_S}{d\zeta_S} \end{aligned} \tag{3.11}$$

κ_I and κ_S are the heat conductivities for ice and snow. The temperature at the bottom of the ice layer is set to the freezing temperature of sea water.

3.3 Turbulence scheme

The original COSMO code was not constructed for large eddy simulations (LES) and the parameterization of a subgrid-scale model was needed (Herzog et al., 2002b). Therefore a 3d prognostic turbulent kinetic energy (TKE) scheme was developed (Herzog et al., 2002a,b). The turbulent fluxes for heat, mass and momentum were implemented. They are specified by a first-order closure through a local gradient and a local turbulence coefficient. The turbulence coefficients for heat and momentum, K_h and K_m, respectively, follow the description of Smagorinsky (1963) and Lilly (1962) (Langhans et al. (2012), Eq. 3.12). K_h and

K_{m} include the effects of buoyancy by including the Richardson number.

$$K_{\mathrm{m}} = c_{\mathrm{s}}^{2}\Lambda^{2}SF_{\mathrm{m}}, \quad K_{\mathrm{h}} = c_{\mathrm{s}}^{2}\Lambda^{2}SF_{\mathrm{h}} \tag{3.12}$$

The Prandtl-Kolomogorov specification is used for the length scale Λ (Herzog et al., 2002b). The Smagorinsky constant c_{s} is assumed to be 0.25 (Herzog et al., 2002b; Langhans et al., 2012). The functions F_{m} and F_{h} depend on the stability, thus on the Richardson number (R_{i}) and on the critical Richardson number ($R_{\mathrm{i_c}}$) specified by Brown et al. (1994) (Langhans et al. (2012), Equation (3.13)).

$$F_{\mathrm{m}} = \sqrt{max(0, 1 - \frac{R_{\mathrm{i}}}{R_{\mathrm{i_c}}})} \tag{3.13}$$

The Brunt-Väisälä frequency (N) is described for saturated air (N_{m}) after Durran and Klemp (1982) and N is a measure of the strength of the buoyancy force. For calculation of R_{i}, N is squared and divided by $S = S_{\mathrm{ij}}$, the strain tensor, which is also squared ($S^2 = 2S_{\mathrm{ij}}S_{\mathrm{ij}}$) (Equation (3.14)).

$$R_{\mathrm{i}} = \begin{cases} \frac{N_{\mathrm{m}}^2}{S^2} & \text{for saturated air} \\ \frac{N^2}{S^2} & \text{for unsaturated air} \end{cases} \tag{3.14}$$

3.4 Model setup

The semi-idealized LES setup has a horizontal size of 6.4 x 6.4 km with a grid spacing of 100 m. The boundary conditions are periodic similar to Paukert and Hoose (2014). The subsidence is described with a linear function from zero at the surface to 0.4125 cm s^{-1} at the initial temperature inversion height (z_{Tin}) and is kept constant above the initial inversion base height. A fixed cloud droplet number concentration (CDNC) and ice crystal number concentration (ICNC) is used because in this thesis no aerosol information was available. The CDNC at each time step and grid point is set to the fixed CDNC if LWC is greater than 0 (Equation (3.15)).

$$\mathrm{LWC} > 0 : \mathrm{CDNC} = \frac{\mathrm{CDNC_{fixed}}}{\rho} \tag{3.15}$$

After a spin-up time of 2 h cloud ice processes are turned on. Thus, the liquid cloud layer has 2 h time to develop without influences of ice microphysical processes.

The ice nucleation rate (N_i) is calculated if the S_i is greater or equal to 0.05 and if a certain amount of LWC is formed in a grid point (Equation (3.16)). Hence, the ICNC is relaxed to the initially set ICNC ($N_{\mathrm{i}0}$) at every time step (Δt) (Morrison et al., 2011; Ovchinnikov et al., 2014).

$$\begin{aligned}
\frac{\partial N_\mathrm{i}}{\partial t} &= max\left(0, \frac{N_{\mathrm{i}0} - N_\mathrm{i}}{\Delta t}\right), \; S_\mathrm{i} \geq 0.05 \quad \text{or} \quad \mathrm{LWC} \geq 0.001\,\mathrm{g\,kg^{-1}} \\
\frac{\partial N_\mathrm{i}}{\partial t} &= 0, \qquad\qquad\qquad\quad S_\mathrm{i} < 0.05 \quad \text{or} \quad \mathrm{LWC} < 0.001\,\mathrm{g\,kg^{-1}}
\end{aligned} \tag{3.16}$$

The approaches of the fixed CDNC and ICNC, the 2 h spin up time and the description of the large scale subsidence are based on the model setup described in Ovchinnikov et al. (2014). The sun zenith angle is constant and set to the time of the start of the simulations for all different simulations, except for the Impact of Biogenic versus Anthropogenic emissions on Clouds and Climate: towards a Holistic UnderStanding (BACCHUS) simulations (Chap. 7). There the sun zenith angle changes with progressive time. The initial temperature, moisture, and wind profiles were taken from radiosonde or dropsonde profiles, depending on the chosen campaign. The domain height is greater than the dropsonde profile height, therefore the ERA-Interim reanalysis data is used to fill the missing data in the initialization profiles (Dee et al., 2011). CDNCs are based on observations and ICNC are estimated from observations and the instrument detection range. The model setup differs in domain top, vertical levels and subsidence description. Specifications for each chapter are described in Tab. 3.2.

Setup of Chapter 4 The simulations of four days during the VERtical Distribution of Ice in Arctic clouds (VERDI) campaign are made with a model domain top at 22 km, divided into 237 vertical levels. The vertical resolution is 15 m until 1005 m and decreases exponentially above 1005 m. For all four simulations four different dropsonde profiles from the VERDI campaign are used to build the initial profiles for the simulations. The profiles are modified and sometimes smoothed to avoid numerical issuses of the model. The number concentration of ice particles was often below or at the detection limit of the Small Ice Detector mark 3 (SID3) instrument during the VERDI campaign. Thus, the fixed ice crystal concentration is assumed to $1\,\mathrm{l^{-1}}$ based on measurements during the Indirect and Semi-Direct Aerosol Campaign (ISDAC) campaign (McFarquhar et al., 2011; Ovchinnikov et al., 2014). The CDNC of the 29 April 2012 simulation is based on the SID3 measurements and therefore set to 75 cm^{-3}. The same instrument is used to set the CDNC of the 15 May 2012 to 100 cm^{-3}. For the simulations of the other two days, the 14 May and 17 May 2012, the CDNC is assumed to be 100 cm^{-3}. The z_{Tin} for the subsidence description is adjusted for each single simulation of the different days (Tab. 3.2).

Table 3.2: Model setup specifications of the different simulations.

Campaign and simulation day	Specifications	
VERDI, 29 April 2012	z_{Tin} = 500 m	CDNC = 75 cm^{-3}, ICNC = 1 l^{-1}
VERDI, 14 May 2012	z_{Tin} = 870 m	CDNC = 100 cm^{-3}, ICNC = 1 l^{-1}
VERDI, 15 May 2012	z_{Tin} = 988 m	CDNC = 100 cm^{-3}, ICNC = 1 l^{-1}
VERDI, 17 May 2012	z_{Tin} = 350 m	CDNC = 100 cm^{-3}, ICNC = 1 l^{-1}
ASCOS, 31 August 2008	z_{Tin} = 960 m	CDNC and ICNC see Tab. 6.1
BACCHUS, 31 August 2008	-	CDNC and ICNC see Tab. 7.1

Setup of Chapter 5 The sensitivity study about the influences of different surface types and different boundary layer (BL) structures are simulated with a domain top of 2 km. The vertical axis is divided into 133 vertical levels. For this low domain height a simple radiation scheme is used. It is a simple longwave radiation parameterization and depends only on the cloud liquid water path (LWP) in the model (e.g. Stevens et al., 2005; Ackerman et al., 2009; Ovchinnikov et al., 2014). The simple parameterization represents the radiative cooling at cloud top due to variations in LWC. The vertical resolution, the fixed CDNC, and ICNC are the same as used in the VERDI campaign simulations (Chapter 4). For this study the 29 April and 15 May 2012 are analyzed and z_{Tin} are specified in Tab. 3.2.

Setup of Chapter 6 For the sensitivity study of the dissipation of an Arctic mixed-phase cloud the height of the model domain is at about 22 km to consider the influences from the atmosphere above the BL on the mixed-phase cloud. The vertical resolution is similar to Chapter 4 and the model height is divided into 237 vertical levels. The initial temperature inversion base is at 960 m and the subsidence profile is adjusted to this height. The model surface is sea ice with a temperature of 271.35 K and an albedo of 70 %. The albedo corresponds to observations during the ASCOS campaign. The subsidence values are consistent with European Centre for Medium-Range Weather Forecasts (ECMWF) reanalysis of the 31 August 2008. A radio sounding from the 31 August 2008 at 5:35 UTC builds the initial moisture and temperature profiles for the simulation. The wind speed and the wind direction are smoothed from 12 km to 22 km. The altered CDNC and ICNC are shown in Tab. 6.1. The CDNC value for the control simulation is estimated from CCN counter measurements on board the ice breaker (Martin et al., 2011). Unfortunately, there were no observations of ice nuclei (IN) concentrations, because they were low or below the detection limit.

Setup of Chapter 7 The analysis of the microphysical rates is part of the BACCHUS project. The model setup is the same as in Chap. 4, except for the subsidence description. The large scale subsidence is described by a linear function which depends on the vertical height and is decreasing from the surface with a factor of $1.5\text{x}10^{-6}$ towards the model top. Specifications of the simulations are defined in Tab. 7.1.

PART II

RESULTS

CHAPTER 4

Case studies of clouds observed during the VERtical Distribution of Ice in Arctic clouds (VERDI) aircraft campaign

This chapter analyses four different simulations of clouds during the VERDI campaign. Within this campaign, aircraft measurements were performed in April and May 2012 over the Beaufort Sea, north of Canada and around Inuvik, located in the Northwest Territories of Canada (Klingebiel et al., 2015) (Figure 4.1). Thirteen flights with the POLAR 5 research aircraft through clouds provide an insight into profiles of different cloud microphysical parameters and meteorological parameters, such as temperature or humidity. Simulations with focus on cloud formation, lifetime, and properties related to the boundary layer (BL) structure of four different flights during the VERDI campaign are conducted using the model setup described in Section 3.4.

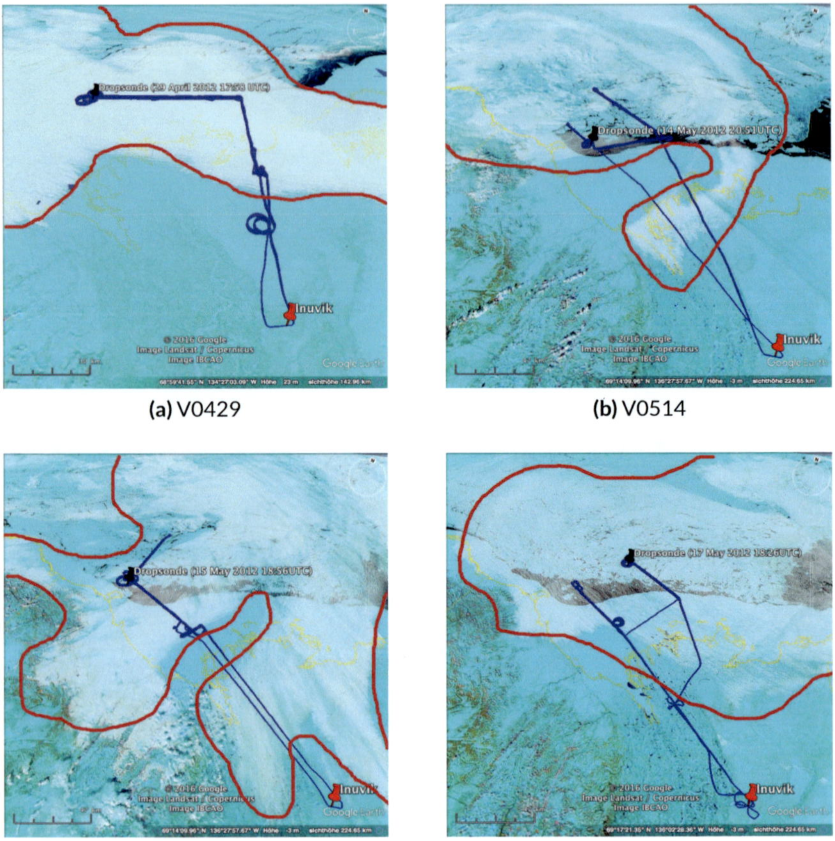

(a) V0429 (b) V0514

(c) V0515 (d) V0517

Figure 4.1: VERDI flight tracks (blue lines) and MODIS images (Aqua; 500 m resolution; bands: 7, 2, 1) are shown for all four simulated flights (https://verdi.fz-juelich.de/ with modifications). Dropsondes are marked with a black pin and the specific date and time. The approximatetly distribution of the cloud decks are indicated with the red lines. The airport Inuvik is marked with a red pin.

4.1 Cloud observed during flight V0429

During the 29 April 2012 several low pressure systems controlled the area of Canada (Bär, 2012). The aircraft observations took place at the southern end of a high pressure system. The pressure gradients were low, and so was the wind

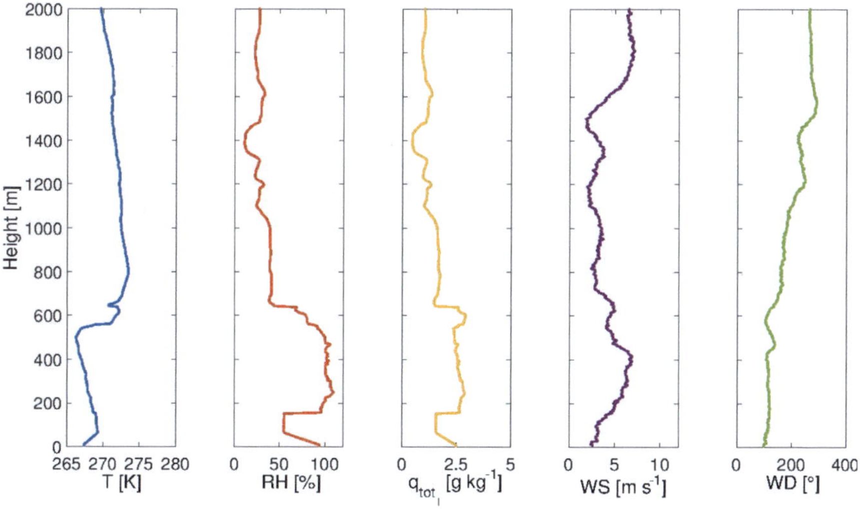

Figure 4.2: V0429 initialization profiles of T, RH, $q_{\mathrm{tot_l}}$, WS, and WD (from left to right).

speed. Clouds of the trough and clouds of the low pressure systems at the surface in the south east of the flight track built a continuous cloud cover (Figure 4.1a). No high clouds were observed and the low level clouds were persistent during the measurement period. The low level cloud cover was spatially and homogeneously distributed. The cloud top was at about 570 m and the cloud base at about 150 m. The initialization profiles of temperature (T), relative humidity (RH), total specific humidity (water vapour and water) ($q_{\mathrm{tot_l}}$), wind speed (WS), and wind direction (WD) (Figure 4.2) are based on dropsonde profiles from 17:58 UTC and measurements on the aircraft. WS and WD are smoothed by a moving average filter to avoid numerical issues in the model. The RH and $q_{\mathrm{tot_l}}$ of the aircraft measurements are taken into account to calculate $q_{\mathrm{tot_l}}$ inside the cloud and set the vertical borders of cloud top and cloud bottom. Figure 4.3 shows time series of liquid water content (LWC) and ice water content (IWC) of the simulation. At the temperature inversion height (at about 550 m) a liquid layer develops. Maximum values of LWC are around $0.25\,\mathrm{g\,kg^{-1}}$ and occur near cloud top (Figure 4.3). The cloud top height after 5 h is at about 590 m and the cloud base height is at about 135 m. The cloud top is raising because the BL mixes and grows in vertical direction. The simulated cloud top agrees well with the observed height at 570 m.

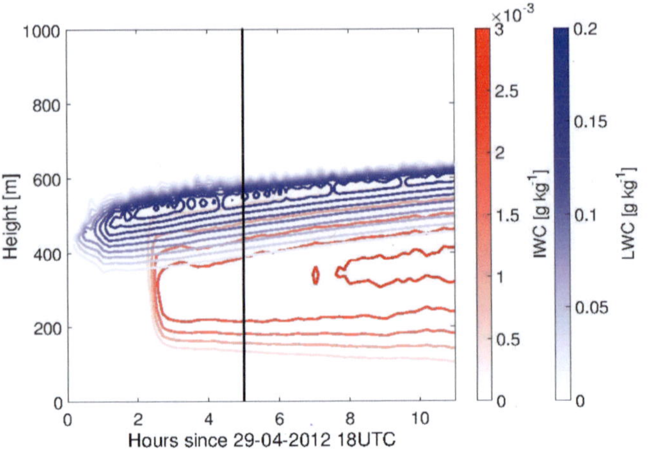

Figure 4.3: Mean values of LWC (blue), IWC (red) of the V0429 simulation during the VERDI campaign. The black line marks 5 h.

The liquid layer is above the ice layer and in the middle of these two layers a mixed layer develops. The ice forms in the liquid layer and the ice particles grow on their way down to the surface. The amount of snow water path (SWP) is very small (Figure 4.4, dashed blue line) and the ice particles do not reach the surface (Figure 4.3, red). Therefore, the precipitation can be ignored, even if some rain drops form because they do not reach the surface (Figure 4.5, dashed blue line). For a closer look at the BL structure, the potential temperature (θ) and equivalent potential temperature (θ_e) profiles are analyzed (Equation (4.1) and Equation (4.2)). In this thesis the θ and θ_e profiles are calculated with following equations:

$$\theta = T \frac{p_0}{p}^{\frac{\mathscr{R}_d}{c_p}} \tag{4.1}$$

$$\theta_e = \theta exp\left(\frac{L_v QV}{c_p T}\right) \tag{4.2}$$

Compared to θ, the equivalent potential temperature accounts for moisture in the atmosphere with the specific humidity (QV) and latent heat of condensation (L_v) ($2.5 \times 10^6\,\mathrm{J\,kg^{-1}}$). In both equation following constants are important: the standard reference pressure (p_0) (1000 h Pa), the specific gas constant of dry

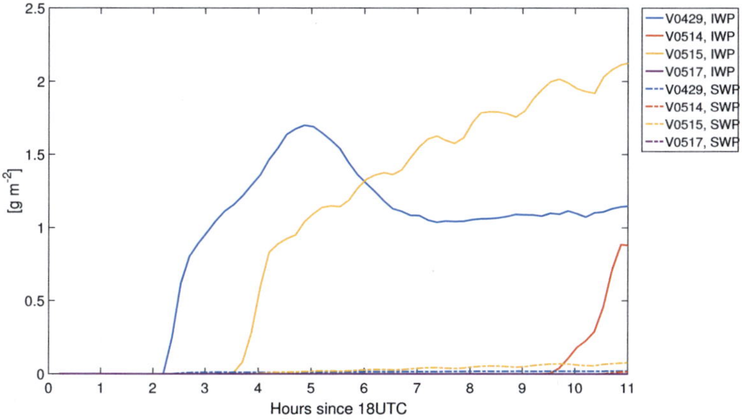

Figure 4.4: The IWP (solid lines) and the SWP (dashed lines) of the four VERDI campaign simulations, V0429 (blue), V0514 (red), V0515 (yellow), and V0517 (purple).

air ($\mathscr{R}_d$) (287.058 J kg^{-1} K^{-1}), and the specific heat at constant pressure (c_p) (1005 J kg^{-1} K^{-1}).

The BL is decoupled from the surface seen by inversion near the surface. The θ profiles show that the BL stays decoupled during the whole simulation time (Figure 4.6a). The stable part of the BL above the decoupled surface layer becomes more and more well mixed. On the other hand, the θ_e profiles show a small inversion at about 180 m, which weakens over the simulation time (Figure 4.6a). This inversion near the surface corresponds to the cloud bottom of the simulated cloud. Both, θ and θ_e, show an inversion height at about 500 m, which increases towards 600 m height over time like the cloud top height. Thus, the whole BL deepens with time due to entrainment through the cloud top.

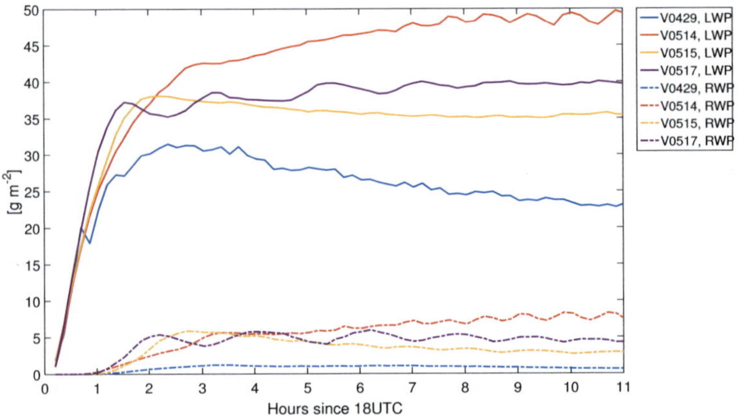

Figure 4.5: The LWP (solid lines) and the RWP (dashed lines) of the four VERDI campaign simulations, V0429 (blue), V0514 (red), V0515 (yellow), and V0517 (purple).

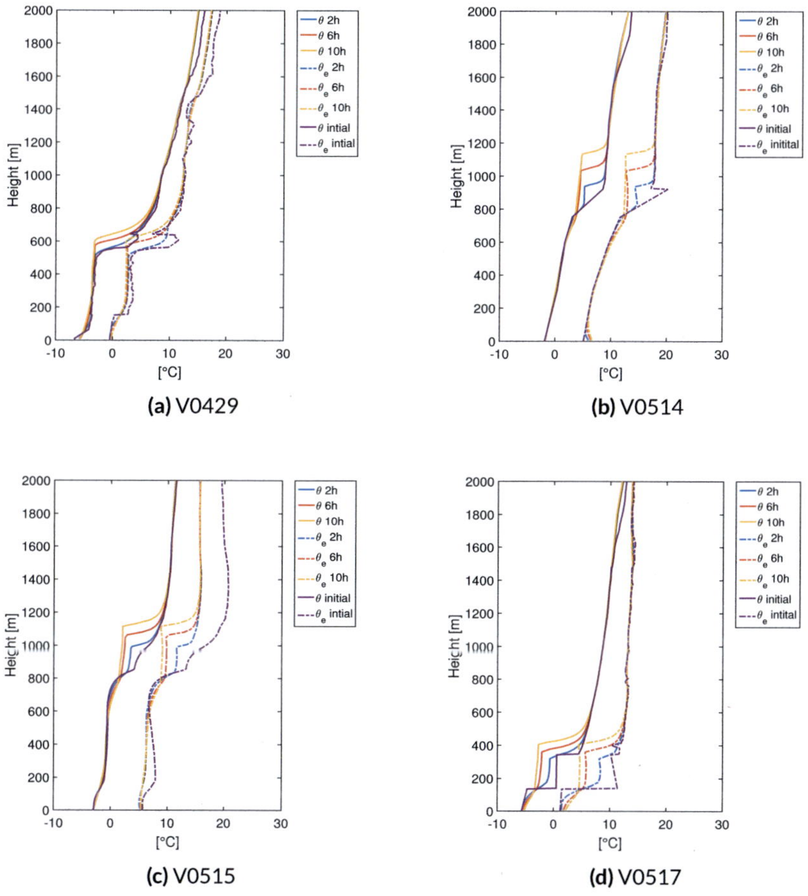

Figure 4.6: θ profiles (solid lines) and θ_e profiles (dashed lines) of the four different simualtions for three different time steps and of the initialization profiles.

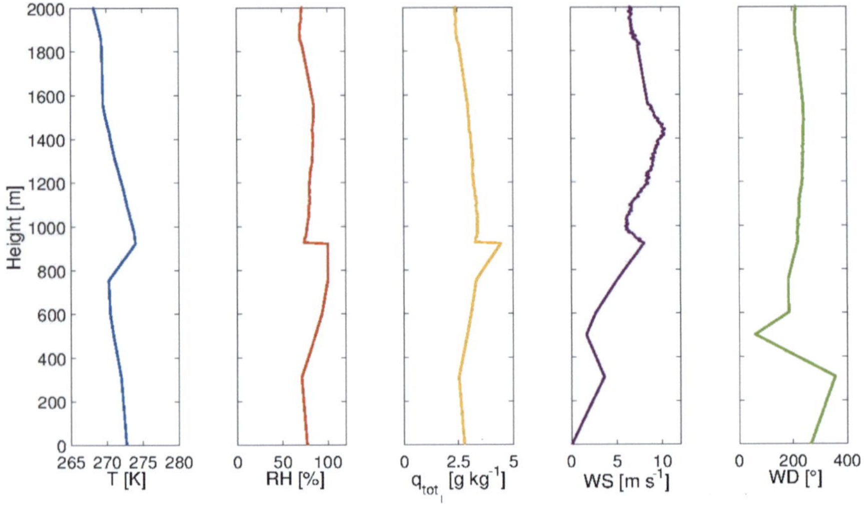

Figure 4.7: V0514 initialization profiles of T, RH, q_{tot_l}, WS, and WD (from left to right).

4.2 Cloud observed during flight V0514

A high pressure area evolved over the whole region of observation during the 14 May 2012. Relatively warm surface temperatures between -2 °C and 12 °C were measured (Bär, 2012). The high pressure area together with the high pressure system at the ground formed an Omega block. South West of the flight track a low pressure system at the surface moved east and dissolved the Omega block during the day. Furthermore, this low pressure system formed clouds and the cold air masses of the high pressure system assists the cloud forming process (Figure 4.1b). The wind at the surface was weak coming from South East and no precipitation was observed. The observed cloud was around 100 to 150 m thick and had a patchy structure. The cloud top height was at about 870 m and the cloud bottom height is at about 640 m. A few higher clouds were also observed during the flight.

The initialization profiles are smoothed by removing the most fluctuating points of the dropsonde profiles because the dropsonde profiles from 20:51 UTC varied a lot and this cause numerical issues (Figure 4.7). The RH inside the cloud is

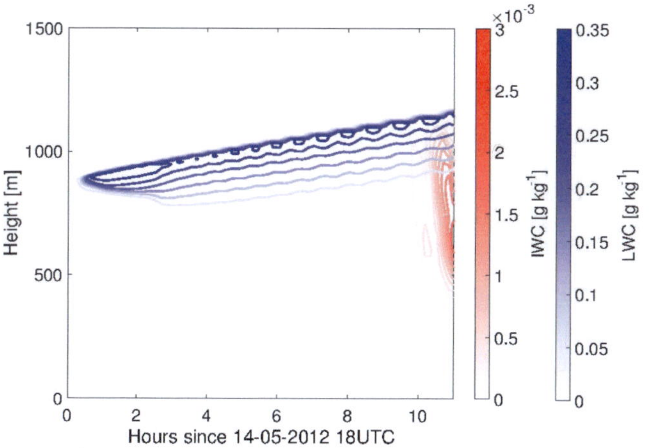

Figure 4.8: Mean values of LWC (blue) and IWC (red) of the V0514 simulation during the VERDI campaign.

estimated from the highest measured RH and from the RH of the similar case V0515 and thus set to 100.2 %. The profiles show a weak temperatures inversion between 800 m and 1000 m. Thus, the simulated cloud develops at the inversion height. The cloud is most of the time liquid, because the temperatures are to high to form ice particles (Figure 4.8 and Figure 4.4, red lines). This fits also to the observations, were no ice was observed. Nevertheless, some ice forms at the end of the simulation at around 10 h after the start of the simulation. This is due to the fact that the ice parametrization is turned on and ice can form if the conditions are favorable. If the supersaturation over ice reaches 0.05 and liquid water exists, ice mass can grow in the model (Equation (3.16)). The temperatures inside the simulated cloud layer are above 0 °C at the beginning. Due to radiative cooling it is getting colder reaching around -5 °C eventually. The liquid cloud is at about 200 m thick, which is comparable to the observations.
The inversion height increases within 4 h by at about 100 m in height (Figure 4.6b). This is also seen in the raising liquid cloud top. The θ profiles show a stable BL from the surface to the inversion height, similar to the θ_e profiles. In addition, the θ_e profiles show a vertical thin unstable surface layer after 2 h of simulation. This layer lasts for the rest of the simulation time and does not show an influence

on the cloud layer. This indicates that the surface stays warm around 0 C° and is becoming more moist compared to the beginning of the simulation.

4.3 Cloud observed during flight V0515

The high pressure area, which was observed the day before weakened (Bär, 2012). At the surface, to the North East of the flight track, a high pressure system was located and in the North West a low pressure system was present. During the flight the cloud cover moved only a little bit from South West over the observation area. However, the cloud sheet got thicker and horizontally wider. Some middle altitude clouds were observed, but the majority were low level clouds (Figure 4.1c). The temperature decreased compared to the temperatures on the 14 May 2012 and ranged between -7 °C and 3 °C at the surface. No precipitation was observed during the flight. The observed cloud top was at about 990 m and only liquid droplets were measured near cloud top (Klingebiel et al., 2015). However, at the bottom of the cloud, at a height at about 670 m, some ice columns and snowflakes were observed. Overall, the Arctic mixed-phase cloud was capped by a warmer and drier atmospheric layer.
The initialization profile for T, $q_{\mathrm{tot}_\mathrm{l}}$, and RH are constructed from aircraft measurements and combined with dropsonde profiles from 18:56 UTC (Figure 4.9). WS and WD are only based on dropsonde profiles. Figure 4.10 shows the simulated cloud. The simulated cloud top is at a height around 1 km, at which also the temperature inversion is located (Figure 4.10 and Figure 4.6c). The maximum LWC is about 0.3 g kg^{-1}, with highest values near cloud top. The RWP shows that some rain drops are formed after 1 h with a maximum value of around 5 g m^{-2} and RWP is decreasing after 3 h (Figure 4.5, yellow dashed line). However, only a small amount of SWP (< 0.1 m^{-2}) is seen (Figure 4.4, yellow dashed line). The IWC reaches maximum values of around 0.003 g kg^{-1} below the liquid cloud layer at approximately 630 m. By comparison of the observed LWC with the 2 h mean of the simulated LWC, the two profiles are in a good agreement. The observed maximum LWC at cloud top is only around 10 m below the simulated maximum LWC (Figure 4.11) (Klingebiel et al., 2015). The simulated liquid

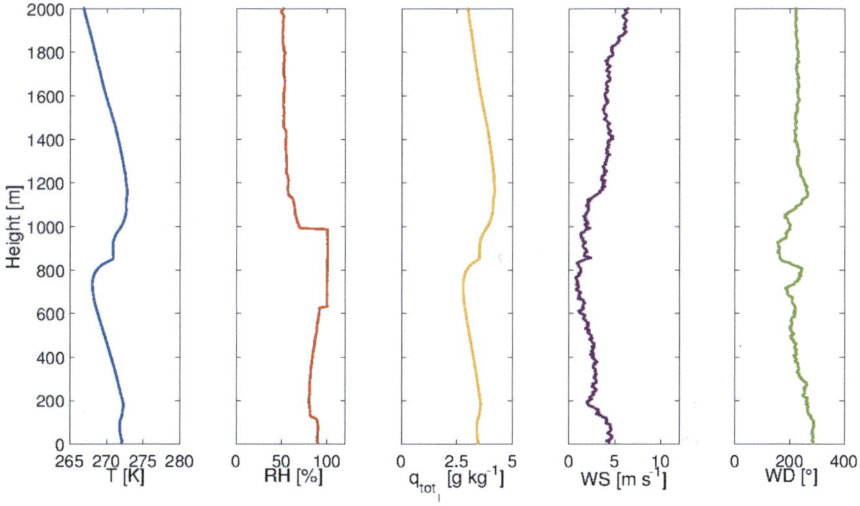

Figure 4.9: V0515 initialization profiles of T, RH, q_{tot_l}, WS, and WD (from left to right).

cloud layer is around 200 m thick, whereas the observed liquid layer extends further down than the simulated one. However, the simulated LWC profile is in the range of the uncertainty of the observed LWC profile in the upper part of the cloud (Figure 4.11).

The θ and also the θ_e profiles show a decoupled surface layer with a vertical thickness of around 100 m (Figure 4.6c). It lasts for the whole simulation time. The inversion height as well as cloud top increase with time from 800 m up to 1100 m after 10 h. Between the decoupled surface layer and the inversion base height at about 800 m, the BL is neutrally stratified. The θ_e profiles show an unstable BL at the beginning of the simulation, which is becoming more and more stable. This limits the growth of the mixing layer towards the surface.

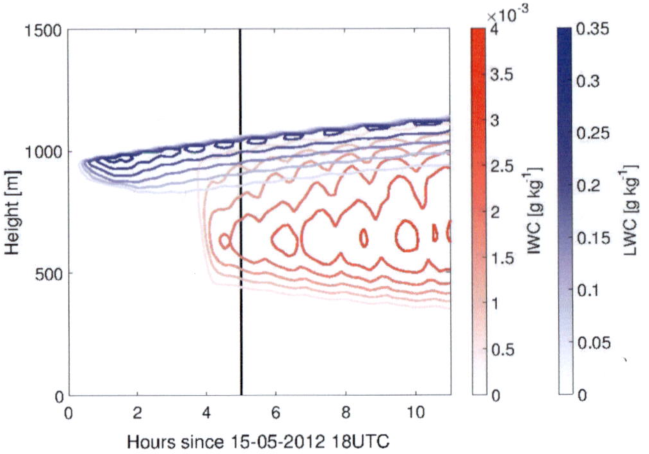

Figure 4.10: Mean values of LWC (blue) and IWC (red) of the V0515 simulation during the VERDI campaign. The black line marks 5 h.

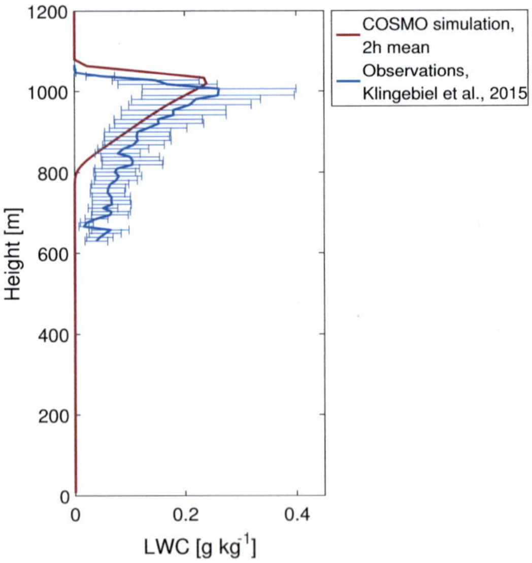

Figure 4.11: 2 h mean (hours 4 to 6) of the simulated LWC profile (red) compared to the observed LWC (with error bars in blue). The error is calculated by error propagation from measured paramters.

4.4 Cloud observed during flight V0517

During the 17 May 2012 a stationary new high pressure area covered the observation region (Bär, 2012). The high pressure system at the surface led to relatively warm temperatures with a minimum temperature at -4 °C and low wind speeds. The whole flight area was covered with a mixed-phase cloud and no precipitation, but fog was observed at the surface (Figure 4.1d). The cloud top was at about 300 m and the thickness of the cloud varied from 200 m to 250 m. The cloud was mostly liquid, still a few ice particles were observed.
The dropsonde profiles from 18:26 UTC are smoothed by removing the most fluctuating points in the profiles because the profiles vary a lot and this causes numerical issues in the model. The initialization profile of temperature shows two inversions, one is at about 190 m and the other is at about 390 m (Figure 4.12). The simulated cloud has a maximum LWC of around 0.25 $g\,kg^{-1}$. The IWC is in the range of 10^{-8} $g\,kg^{-1}$ and therefore negligible (Figure 4.13). Additionally, almost no SWP is seen (Figure 4.4, purple dashed line). Thus, no snow is formed. The liquid cloud top rises from at about 300 m to at about 420 m and at the same time the inversion height is rising (Figure 4.6d). The mostly liquid cloud is stable for the whole 11 h of simulation.
The θ and θ_e profiles show an unstable surface layer (Figure 4.6d). Furthermore, between the two inversions, where the liquid cloud layer is located, the BL is well mixed. The inversion near the surface below 200 m persists during the whole simulation and because of a RH below 100 % no liquid cloud water forms and indicates fog, which was observed during the flight.

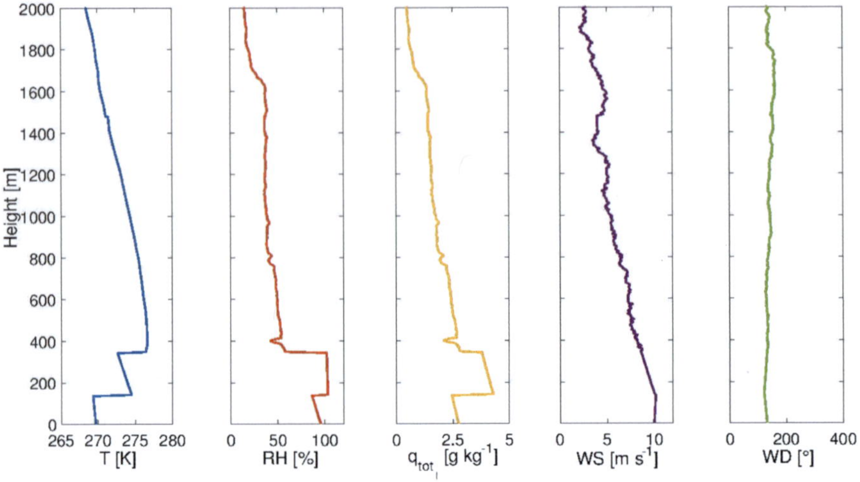

Figure 4.12: V0517 initialization profiles of T, RH, q_{tot_l}, WS, and WD (from left to right).

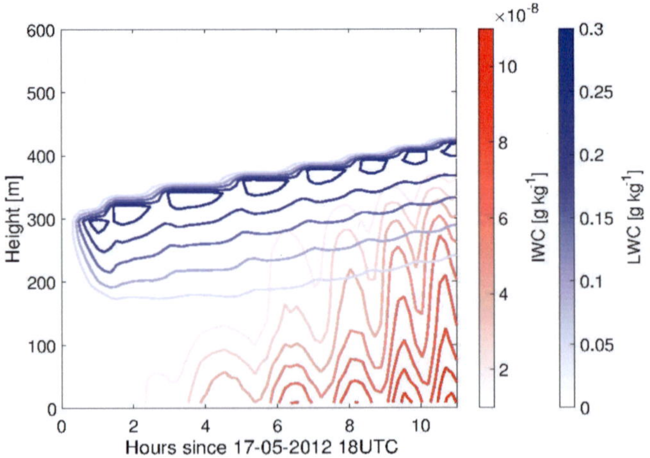

Figure 4.13: Mean values of LWC (blue) and IWC (red) of the V0517 simulation during the VERDI campaign.

4.5 Analyzing the droplet size near cloud top

The Small Ice Detector mark 3 (SID3) is built to use in an aircraft and was mounted below the wing of the Polar 5 aircraft during the VERDI campaign. The instrument captures the spatial light scattering patterns of individual cloud and ice particles (Vochezer et al., 2016). The measurements during the VERDI campaign were done by Paul Vochezer and Martin Schnaiter (Institute of Meteorology and Climate Research, Department Atmospheric Aerosol Research (IMK-AAF), KIT). An intensified charged-coupled device camera records the high resolution images of the scattering patterns. The shutter release of the camera activates through a signal of a trigger detector. Thus, pictures of ice particles and droplets can be analyzed. Additionally, the particle number size distribution and particle number concentration can be derived from the pulse intensities and trigger count rate (Vochezer et al., 2016).
The droplet diameter derived from the simulated LWC and the number density of cloud droplets (QNC) is shown in Figure 4.14 for V0429 and in Figure 4.15 for V0515 (Equation (3.3) - Equation (3.5)). Both cases show the domain averaged mass diameter at 5 h after the simulation start. The droplet diameter increases with increasing height and towards the highest values of the LWC (Figure 4.3 and Figure 4.10, blue). The same is seen by the SID3 (Figure 4.14 and Figure 4.15, blue). The droplets are growing by condensation on their way trough the liquid layer of the cloud. At cloud top the droplets evaporate and get smaller because of, for instance, entrainment of dryer air above the cloud. The linear fits through the data points show a slightly difference in the slope. For V0429 the slope is steeper in the measurements, while for V0515 the droplet diameter increase is similar to the observations. Comparing the droplet growth through the cloud layer with the wet adiabatic growth of droplets assuming a monomodal droplet size distribution, the slope of the V0429 simulation is closer to the slope of the wet adiabatic profile than the slope of the V0515 simulation. Both slopes from the simulations and observations deviate from the wet adiabatic slope because of influences from for instance entrainment processes and ice crystal growth. The wet adiabatic profiles assuming a Γ-distribution of the droplet size diameters increase faster than the monomodal distribution and is closer to the simulations because the model prescribed a Γ-distribution of the droplet diameters.

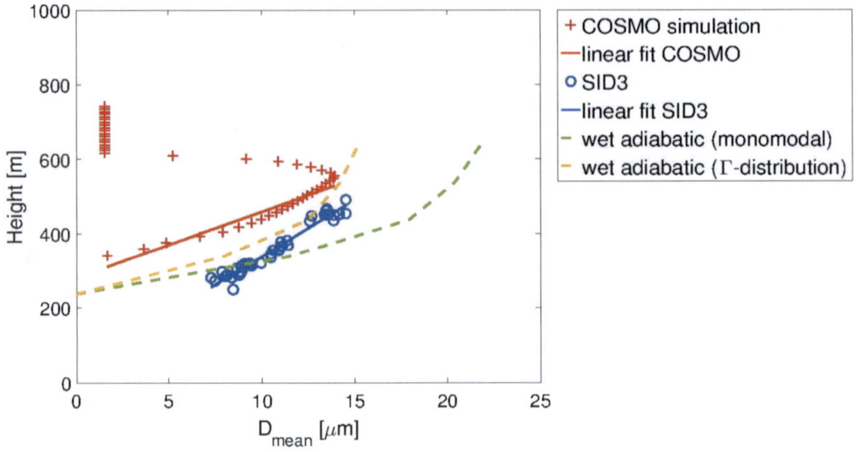

Figure 4.14: Mean droplet diameter after 5 h of simualtion (red) and diameter measured by the SID3 (courtesy of P. Vochezer) (blue) for V0429. Linear fits through the measured and simulated droplet diameter up to the maxium diameter are shown by red and blue lines. The wet adiabatic diamter profiles is calculated with the wet adiabatic LWC profile and assuming a monomodal droplet size distribution (green) and a Γ-distribution (yellow), and a QNC of 75 cm^{-3}.

In general, the simulated mean droplet diameter is smaller then the observed mean droplet diameter. The smaller simulated droplet diameter is also apparent in the droplet size distribution at cloud top (Fig. 4.16). The droplet size distribution is measured with the cloud combination probe (CCP) (Klingebiel et al., 2015). The COSMO model has a prescribed shape of the size distribution of droplets, which indicates to be different to the real droplet size distribution. The V0429 simulation has less droplets in the size range of 8 to 10 μm than the V0515 simulation. This could be due to the colder temperatures in the BL during V0429 compared to V0515 (Figure 4.6a and Figure 4.6c). Thus, less water condenses in V0429 (Figure 4.5, yellow and blue solid lines).

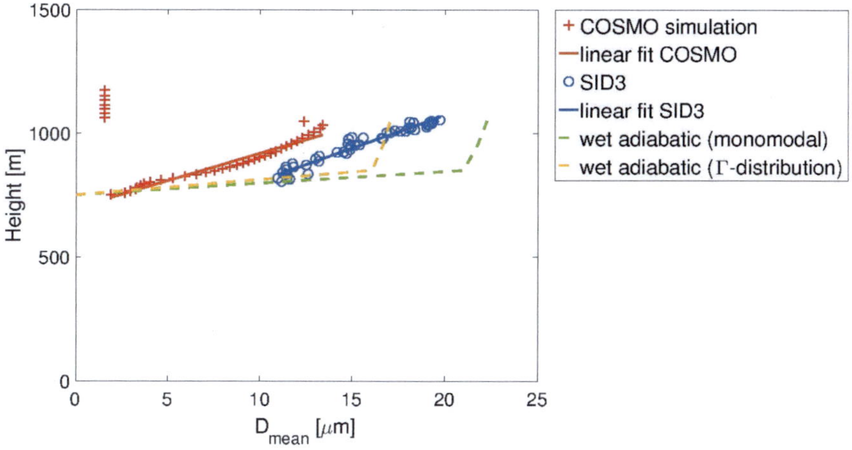

Figure 4.15: Mean droplet diameter after 5 h of simualtion (red) and diameter measured by the SID3 (P. Vochezer) (blue) for V0515. Linear fits through the measured and simulated droplet diameter are shown by red and blue lines. The wet adiabatic diamter profiles is calculated with the wet adiabatic LWC profile and assuming a monomodal droplet size distribution (green) and a Γ-distribution (yellow), and a QNC of $100\,\text{cm}^{-3}$.

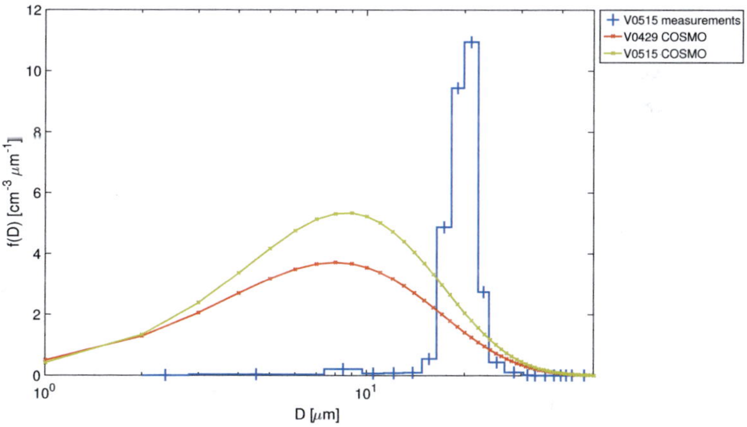

Figure 4.16: The droplet size distribution is measured with the CCP for V0515 at a height of 1008 m (blue) (Klingebiel et al., 2015). Simulated droplet size distribution at cloud top height after 5 h of simualtion, for V0429 at a height around 570 m and for V0515 at a height around 1007 m (red and green, respectively).

4.6 Analyzing the cloud top structure

The horizontal cloud top structure is spatially and temporally variable. Roll convection and cell convection are observed for instance at cold air outbreaks over the Arctic sea (Brümmer, 1999). These two types of cloud structures differ by its orientation. The roll structures have a clearly direction while the cell structures have not. Clouds affect the surface energy budget and changes in cloud properties such as changes in the horizontal structures at cloud top may influence the surface energy budget and the cloud-radiation feedback. The cloud top structure can be observed from airborne nadir observations. However, the entrainment at cloud top can play a role in changes of the cloud top structure. In this section, the entrainment zone of the four simulated mixed-phase clouds during VERDI as well as the cloud top structure are analyzed. Additionally, a short comparison to the measurements from the imaging spectrometer AisaEAGLE is conducted.

4.6.1 The entrainment zone

To investigate the entrainment at cloud top the stability of the BL (Figure 4.6) and hence the inversion height are of interest. The inversion height is derived from the gradients of θ and $q_{\mathrm{tot_l}}$, both become maximal if the inversion height is reached (Figure 4.17a). To describe the depth of the entrainment zone the ratio between the domain averaged LWC and the maximum of LWC in each vertical level is calculated for all four VERDI simulations (Figure 4.18) (after e.g. Burnet and Brenguier, 2007). This ratio shows how far the LWC detrains into the atmosphere above the cloud top. Thus, it is seen how deep the entrainment zone is. The depth ranges from about 100 m for the V0515 simulation to about 200 m for the V0517 simulation at 6 h of simulation time. A dependency on the height of the inversion of the depth of the entrainment zone is seen (Figure 4.17a). The lower the mixed-phase cloud is, means the lower the inversion height is, the deeper the entrainment zone is. The simulations of V0514 and V0515 reach inversion heights in the range of around 920 m and 1150 m, whereas for the other two simulations the inversion heights are more than 500 m lower. Furthermore, these two simulations of V0429 and V0517 have an entrainment depth of around 150 m and 200 m, respectively. The entrainment depth is more than 50 % of the cloud top height in V0517 and around 25 % of the cloud top height of V0429.

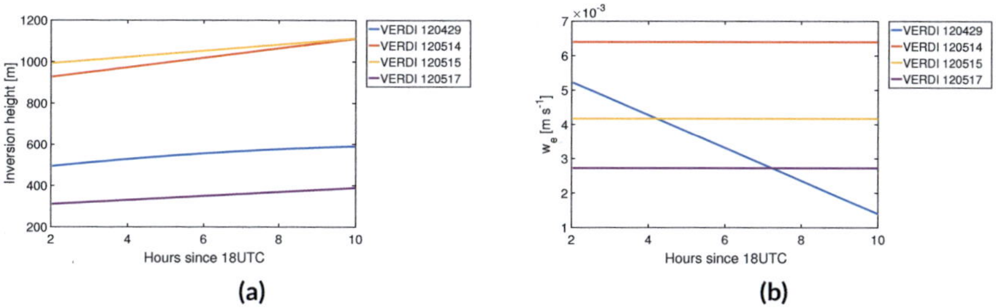

Figure 4.17: The simulated z_i is shown on the left side derived from θ gradients and q_{tot_l} gradients **(a)**. The w_e is derived from z_i **(b)**. Both paramters are shown for all four VERDI campaign simulations (V0429,V0514, V0515, and V0517, colored lines).

Additionally, the evolution of the inversion height in time shows that in all four cases the inversions height is increasing (Figure 4.17a). It follows, that the entrainment zone is moving upward during the course of the simulation. The entrainment velocity (w_e) is derived from the time series of inversion height (z_i) and represents a mean entrainment velocity at the cloud top zone (Equation (4.3)).

$$w_e = \frac{\Delta z_i}{\Delta t} \tag{4.3}$$

The inversion height of the V0429 simulation does not increase as fast as the other simulations and this leads to a decreasing w_e, while during the other simulations w_e stays constant over time (Figure 4.17b).

4.6.2 The simulated cloud top structure

The LWP structures show differences in the cloud top structure for the different simulated clouds (Figure 4.19). Fields of low and high LWP are more structured, for instance in V0429 (Figure 4.19a), than in the other simulations. Moreover, the size of aereas with low and high LWP vary a lot. The wind directions show only a minor influence on the direction of the cloud top structures (Table 4.1). The wind direction do not change a lot during the simulations because of the periodic boundary conditions, which are set in the model. In V0515 the wind is coming from the South and the cloud top structure seems to have a South-North direction (Figure 4.19c). A comparable behavior is seen in the V0517 simulation, in which the wind is coming from South-South-East. The South-West wind direction in V0514 influences the cloud top structure by aligning the structure in a diagonal way. In contrast to that, the cloud top structure of V0429 shows a unclear direction, hence the wind direction in V0514 play a minor role in this simulation. In all four simulations the wind speeds near cloud top are around 5 $\mathrm{m\,s^{-1}}$ and thus the influence on the cloud top structures is similar in all four simulations.

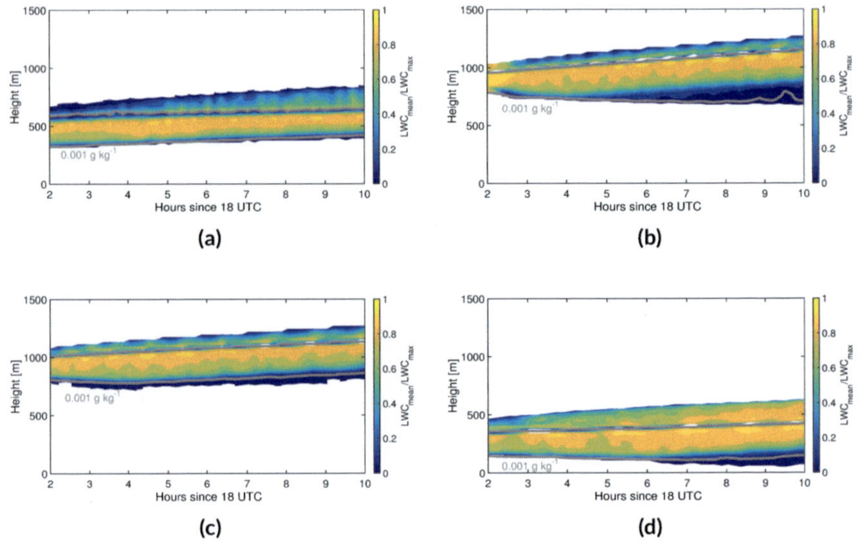

Figure 4.18: Ratio of the horizontal domain mean of LWC and the maximum LWC in each vertical level for the simulations of the clouds of **(a)** V0429, **(b)** V0514, **(c)** V0515, and **(d)** V0517. The grey line marks a LWC of 0.001 g $\mathrm{kg^{-1}}$ indicating the boundary of the liquid cloud layer.

Table 4.1: The initial WD at the cloud top of the four VERDI simulations.

Simulation	Height	WD
V0429	around 600 m	around 111 °
V0514	around 1000 m	around 220 °
V0515	around 1050 m	around 181 °
V0517	around 360 m	around 134 °

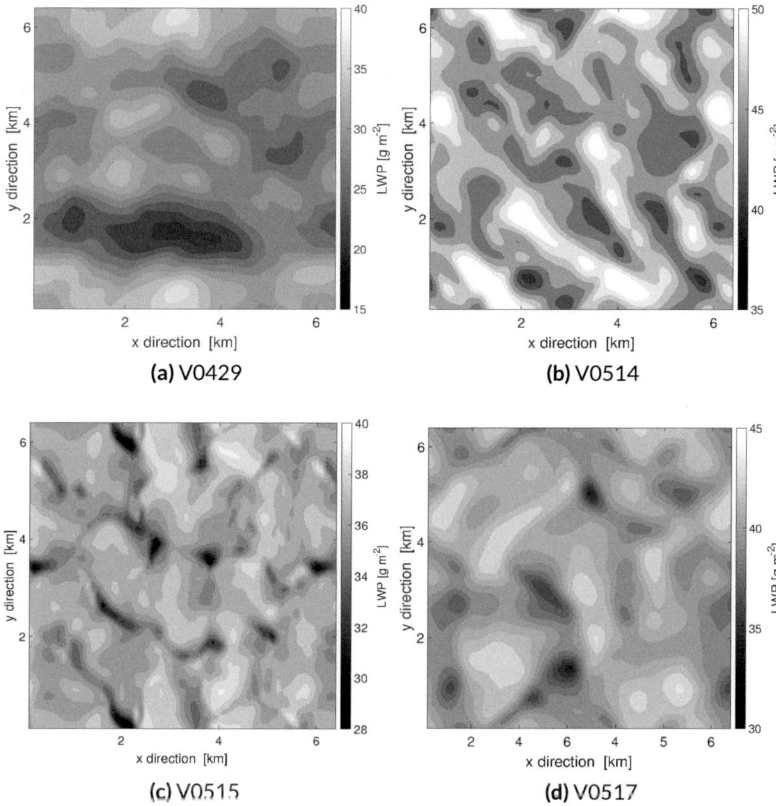

Figure 4.19: The LWP after 5 h 22 min in the simulations of V0429 **(a)**, V0514 **(b)**, V0515 **(c)**, and V0517 **(d)**. Please note the color scale is different in each figure.

4.6.2.1 One-dimensional parameter

The cloud top structure can be quantified and analyzed with one-dimensional inhomogeneity parameters. In this study, the ratio of the standard deviation to the mean ρ_l and the one-dimensional inhomogeneity parameter of l (S_l) are calculated

(Equation (4.4) and Equation (4.5)). For the measurements l is the cloud optical thickness (τ) and for the simulations l is replaced by the LWP. τ depends on the integral of the LWC, thus direct on the LWP and on the effective radius (r_{eff}) of the cloud (Equation (4.6)).

$$\rho_{\mathrm{l}} = \frac{\sigma_{\mathrm{l}}}{\bar{l}} \tag{4.4}$$

$$S_{\mathrm{l}} = \frac{\sqrt{ln(\rho_{\mathrm{l}}^2 + 1)}}{ln 10} \tag{4.5}$$

$$\tau = \int_{z=0}^{h} \frac{3\,\mathrm{LWC}}{2\,\rho_{\mathrm{w}}\,r_{\mathrm{eff}}} \mathrm{dz} \tag{4.6}$$

The data are provided by the Leipzig Institute for Meteorology, University of Leipzig (Schäfer et al., 2017). ρ_{l} is a normalized measure of the cloud inhomogeneity and depends on the standard deviation of l (σ_{l}) and the averaged l ($\bar{l}$) (Davis et al., 1999; Szczap et al., 2000; Schäfer et al., 2017). If ρ_{l} is zero the cloud is homogeneous and if ρ_{l} is greater than zero the cloud top structure is inhomogeneous. However, ρ_{l} can be greater than one and is a qualitative measure of cloud inhomogeneity. S_{l} is a non-linear but monotonous function of ρ_{l}, if l is log-normal distributed in frequency (Davis et al., 1999; Szczap et al., 2000; Schäfer et al., 2017). The logarithm of τ is approximately linear to the reflected or transmitted radiance. For this, τ has to be in the range of around 3 to 30 and $\log\tau$ between 0.5 and 1.5. The higher S_{l} the more inhomogeneous the structure is. Thus, S_{l} is a quantitative measure of cloud inhomogeneities.

The one-dimensional inhomogeneity parameter of LWP (ρ_{LWP}) of all four simulations stays almost constant during time and varations in ρ_{LWP} are only seen in the V0429 simulation (Figure 4.20). This is also the simulation with the highest ρ_{LWP} and the highest one-dimensional inhomogeneity parameter of LWP (S_{LWP}). Hence, this cloud top structure is more inhomogeneous than the other three cloud top structures. The cloud top structures of V0514, V0515, and V0517 are inhomogeneous too, compared to V0429 the ρ_{LWP} and S_{LWP} are lower. ρ_{LWP} is around 0.06 and S_{LWP} ranges between 0.02 and 0.03 (Table 4.2). In general, all four cloud top structure are inhomogeneous and the cloud top structure of the V0429 simulation is the most inhomogeneous one.

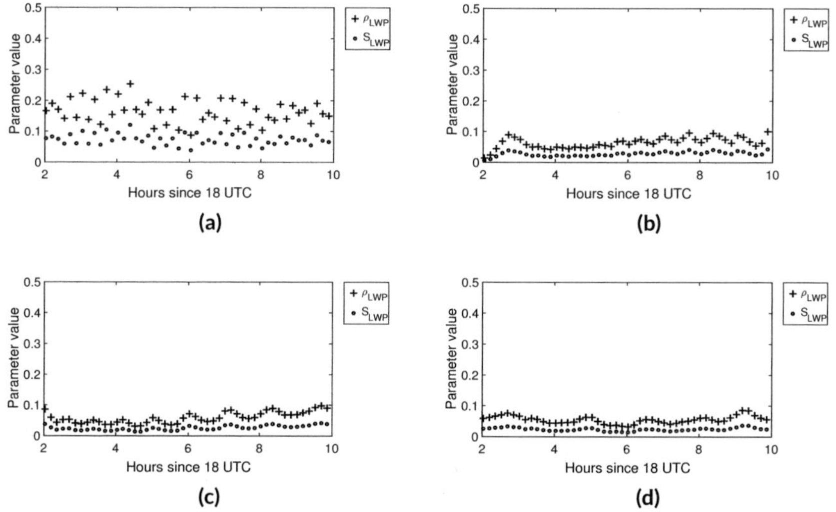

Figure 4.20: The one-dimensional inhomogeneity parameter ρ_{LWP} and S_{LWP} are shown for V0429 **(a)**, V0514 **(b)**, V0515 **(c)**, and **(d)** V0517.

Table 4.2: One-dimensional parameters, ρ_{LWP}, ρ_{τ}, S_{LWP}, S_{τ}, and decorrelation lengths of the four different simulations and of the observations. The decorrelation lengths of the simulations are interpolated to a 10 m gridspace, spatially and temporally averaged over 2 h to 10 h. The parameters from the observation are calculated for the retrived field of τ (Schäfer et al., 2017)

Case	Model simulations						Observations	
	$\bar{\rho}_{\mathrm{LWP}}$	$\sigma_{\rho_{\mathrm{LWP}}}$	$\bar{S}_{\mathrm{LWP}}$	$\sigma_{S_{\mathrm{LWP}}}$	$\bar{\xi}_{\mathrm{LWP},L_x}$ [m]	$\bar{\xi}_{\mathrm{LWP},L_y}$ [m]	ρ_{τ}	S_{τ}
V0429	0.19	0.12	0.08	0.05	501.6	593.3	-	-
V0514	0.06	0.03	0.03	0.01	315.0	334.0	0.19	0.08
V0515	0.06	0.03	0.02	0.01	233.5	293.0	0.18	0.08
V0517	0.06	0.02	0.03	0.01	374.5	385.0	0.16	0.08

4.6.2.2 Spatial autocorrelation function

The two-dimensional field of measurements and simulations can be also used to calculate the spatial autocorrelation function of l (P_l) (Equation (4.7)) (Schäfer et al., 2017). P_l is calculated for both horizontal directions, x and y (compare

Figure 4.19), and depends on the parameter value l (τ or LWP) which is given at the grid points x_{j}, y_{k}.

$$P_{\mathrm{l}}(L_{\mathrm{x}},L_{\mathrm{y}})=\frac{\sum_{\mathrm{j}}^{\mathrm{n}}\sum_{\mathrm{k+1}}^{\mathrm{m}}\left[l(x_{\mathrm{j}}+L_{\mathrm{x}},y_{\mathrm{k}}+L_{\mathrm{y}})-\bar{l}\right]\cdot\left[l(x_{\mathrm{j}},y_{\mathrm{k}})-\bar{l}\right]}{\sum_{\mathrm{j}}^{\mathrm{n}}\sum_{\mathrm{k+1}}^{\mathrm{m}}\left[l(x_{\mathrm{j}},y_{\mathrm{k}})-\bar{l}\right]^{2}} \tag{4.7}$$

L_{x} or L_{y} describes the pixel lag and is thus the distance between the spatial correlation. The number of pixels n and m of the two-dimensional field set the maximum pixel lags, L_{x} and L_{y}. In case of a perfect correlation, P_{l} is -1 or 1, and $P_{\mathrm{l}} = 0$ indicates no correlation. The autocorrelation function is a measure of self similarities and in the current study the similarities in spatial shifts are analyzed. Because the degree of correlation is of interest, the sign is not important and to avoid misunderstandings P_{l} is squared. With the help of the decorrelation length of l (ξ_{l}) one can quantify the length scale (in m) of the typical size of an individual cloud parcel. The more homogeneous the structure is the greater is ξ_{l}. With the squared P_{l} one can calculate ξ_{l} in both horizontal directions (Equation (4.8)). For the simulations the decorrelation lengths are interpolated to a 10 m grid space and spatially and temporally averaged over 2 h to 10 h.

$$P_{\mathrm{l}}^{2}(\xi_{\mathrm{l}})=\frac{1}{e^{2}} \tag{4.8}$$

The squared one-dimensional autocorrelation functions shows the decorrelation in one direction (x or y) and is calculated for every time step (Figure 4.21, grey lines). The decorrelation length of LWP (ξ_{LWP}) is calculated from the time averaged $P^{2}(L_{\mathrm{y}})$ or $P^{2}(\mathrm{L_{x}})$ (Figure 4.21, red lines).
$P^{2}(\mathrm{L_{y}})$ or $P^{2}(\mathrm{L_{x}})$ differ from simulation to simulation in the x- and y-directions (Figure 4.21). The most homogeneous cloud structure is seen in the V0429 simulation. Here, the ξ_{LWP} in y direction is the highest and around 593 m (Table 4.2). The least homogeneous structure derived from V0515, in which the ξ_{LWP} in x-direction is around 233 m (Figure 4.21f). The remaining two simulations, V0514 and V0517, have ξ_{LWP}s in x-direction of 315 m and 374.5 m, respectively. The ξ_{LWP}s in y-direction are around 334 m and 385 m. Hence, the difference between the two ξ_{LWP}s in the two directions is less than 15 m. Thus, the structures are more symmetric than the the structures from the other two cases. The difference of V0429 and V0515 in the ξ_{LWP}s range between 60 m and 90 m, and indicates that the structure is more directed into the y direction (Table 4.2).

The autocorrelation of a single line in one direction is calculated and it is not taken into account, that the structures can be a combination of the two horizontal directions. Therefore, the two-dimensional fields of $P(L_x, L_y)^2_{\mathrm{LWP}}$ are calculated. Now the whole field is shifted to itself, based on the center of the field, to calculate the autocorrelation. $P(L_x, L_y)^2_{\mathrm{LWP}}$ show a more homogeneous structure in V0429 and V0517, because the field with the same value of $P(L_x, L_y)^2_{\mathrm{LWP}}$ is larger than the fields in the other simulations (Figure 4.22). The field of $P(L_x, L_y)^2_{\mathrm{LWP}}$ of V0429 is horizontally directed, while the fields of V0514 and V0515 are only low directed in South-East to North-East direction. Furthermore, the field of V0517 does not show a specific direction. Figure 4.22b and Figure 4.22c show the evolution of the cloud top structure during two days of the campaign. The cloud layer is probably the same for these two days and it evolves from an initial more homogeneous structure (V0514) to a less homogeneous one (V0515), both with a slightly oriented structure. Finally, on V0517 the cloud top structure changes to a more homogeneous structure with a symmetric distribution.

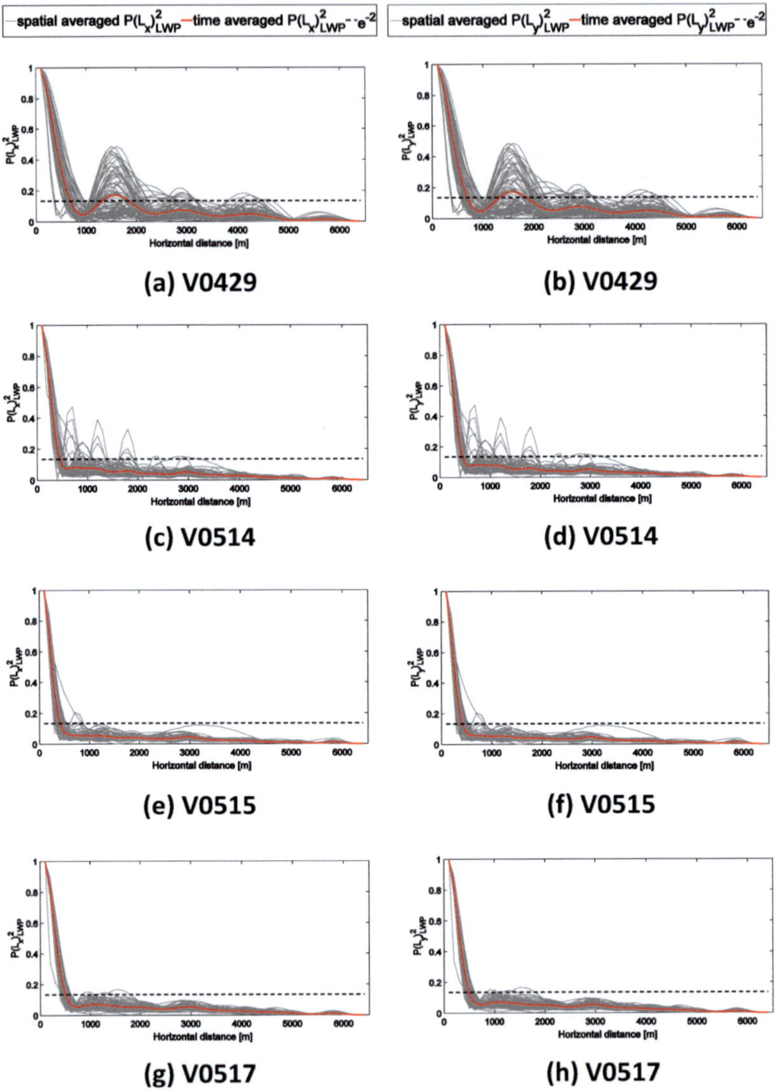

Figure 4.21: The one-dimensional autocorrelation functions in x direction, $P(L_{\mathrm{x}})^2_{\mathrm{LWP}}$ (left), and in y direction, $P(L_{\mathrm{y}})^2_{\mathrm{LWP}}$ (right), at all time steps (grey). The temporal mean of $P(L_{\mathrm{x}})^2_{\mathrm{LWP}}$ and $P(L_{\mathrm{y}})^2_{\mathrm{LWP}}$ is shown in red. V0429 (**a** and **b**), V0514 (**c** and **d**), V0515 (**e** and **f**), and V0517 (**g** and **h**).

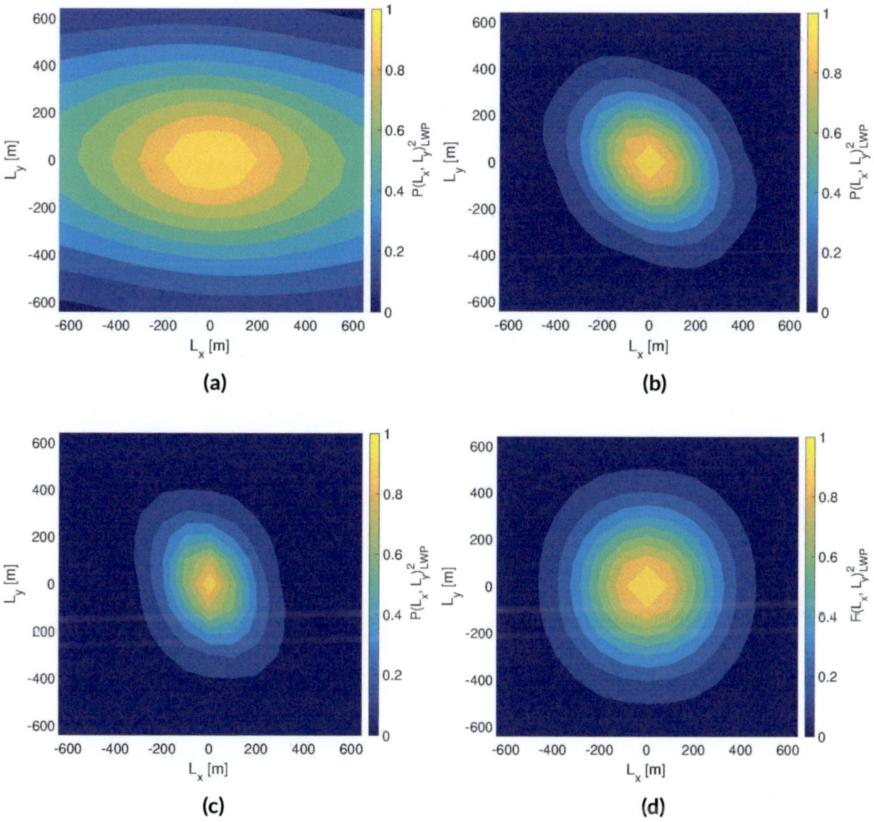

Figure 4.22: The time averaged two-dimensional autocorrelation functions $P(L_{\mathrm{x}}, L_{\mathrm{y}})^2_{\mathrm{LWP}}$ (colours) are shown for V0429 **(a)**, V0514 **(b)**, V0515 **(c)**, and **(d)** V0517.

4.6.3 Comparison of the simulated and observed cloud top structure

During the VERDI campaign the imaging spectrometer AisaEAGLE was mounted on the Polar 5 research aircraft and measured upward solar spectral radiance reflected from the cloud top of Arctic mixed-phase clouds in the wavelength range from 400 nm to 970 nm (Schäfer et al., 2013, 2015). The spectrometer has a single line sensor with a field of view of 37° and 1024 spatial pixels. Because the instrument was mounted below the aircraft, AisaEAGLE made 2D scans of a cloud scene. Using radiative transfer simulations horizontal fields of τ were retrieved from the measurements.
Three of the four simulations can be compared to the derived τ fields of the observations (Table 4.2). The AsiaEAGLE instrument measures in a higher resolution (on average 3 m × 2.7 m) than the simulations (100 m × 100 m) are performed. Hence, all parameter values of the simulations are smaller than the parameters derived from observations. ρ_τ and S_τ only vary slightly between the three observations similarly to ρ_{LWP} and S_{LWP}. Thus, the observed and simulated structures are all inhomogeneous.

4.7 Discussion and Conclusions

In conclusion, this chapter shows results of four different simulations of four different Arctic mixed-phase clouds observed during the VERDI campaign. The four mixed-phase clouds have differences in LWC and in IWC. Besides that, the cloud top heights ranges from around 300 m over 500 m to 1 km and also the cloud layer depth varies from case to case. The COSMO model simulated in all four cases a mixed-phase cloud which is persistent during the whole simulation time. To evaluate the model performance, different aspects of the simulated Arctic mixed-phase clouds are compared to observations. First, the LWC of the V0515 is analyzed. The simulated maximum LWC is in the range of the uncertainties of the observed LWC around 0.2 g kg^{-1} near cloud top (Klingebiel et al., 2015). Furthermore, the liquid cloud top height of the simulation is only around 20 m higher than the observed cloud top and the observed liquid cloud layer is approximately 200 m thicker than the simulated cloud layer.

Second, the droplet sizes in the cloud layer and at cloud top of V0429 and V0515 are investigated (Section 4.5). The droplet sizes are smaller in the V0429 simulation compared to the observations from the SID3 instrument. A better agreement with observations is seen for the V0515 case. Although the droplet sizes are smaller than observed, the entrainment derived from the vertically increasing droplet sizes, appears to be realistic. The simulations show in both cases (V0429 and V0515) that the vertical increase in droplet sizes determined by condensational growth and evaporation of the cloud droplets, due to entrainment at cloud top or to the Wegener-Bergeron-Findeisen (WBF) process agrees to the observed profile of droplet size. Compared to the wet adiabatic droplet profiles the deviation from these profiles is similar in the observed and in the simulated droplet diameter profiles. The observed shape of the droplet size distribution at cloud top of the observations is surprisingly narrow and differs from the simulations. The simulations show a broader droplet size distribution with a maximum of QNC at around 8 μm, while the observations show a maximum of QNC at around 20 μm. As the shape is prescribed in the two-moment scheme, processes leading to the narrowing of the distribution can not be simulated accurately.
Finally, the entrainment rate at the cloud top and the cloud top structure of the simulations are analyzed. In a study by Gerber et al. (2013) entrainment velocitys (w_es) of stratocumulus clouds over the west coast of California were calculated from aircraft observations. The w_es ranges for the specific observed cases between 1.4 and 27.2 mm s^{-1}. The w_es of the four VERDI simulations ranges from around 3 mm s^{-1} for the lowest cloud (V0517) to around 6 mm s^{-1} for the highest cloud (V0514) and thus are in the range of observed stratocumulus clouds. The cloud top structure is compared to the observations from the AisaEAGLE instrument. For this, fields of LWP of the simulations and fields of τ are studied to describe the cloud top structure. The general character of all four simulated cloud tops is according to the inhomogeneity parameters ρ_{LWP} and S_{LWP} inhomogeneous. The same result is found for the character of the three observed cloud top structures (V0514, V0515, and V0517).

Addressing the research question raised in Chapter 1, the results presented here suggest to draw following conclusions:

How well does the model represent an Arctic mixed-phase cloud? How can different observations be used to evaluate the simulations?

- In comparison with measurements the COSMO model simulates comparable Arctic mixed-phase clouds and fits well with the observed LWC and cloud top heights.
- The simulated cloud droplet sizes are smaller than the observed droplet sizes and the prescribed shape of the droplet size distribution is broader than the observed surprisingly narrow distribution.
- The general characteristics of the simulated cloud top structures are inhomogeneous and agree with the characteristics of the observed cloud top structures derived from AisaEAGLE measurements.
- A multitude of observations contributed to validation of the COSMO model.

CHAPTER 5

Influence of surface heterogeneities on Arctic mixed-phase clouds

In the following sensitivity study, the surface type is changed from sea ice to an open lead or sea water surface (Figure 5.1). The open lead area takes 12.5 % of the whole surface area. Thus, simulations with three different surfaces are conducted for two different cases during the VERtical Distribution of Ice in Arctic clouds (VERDI) campaign (flight on 29 April 2012 during VERDI (V0429) and flight on 15 May 2012 during VERDI (V0515)). The wind direction (WD) at the surface is around 100 ° for V0429, hence from East-South-East and the open lead is almost perpendicular to the WD. In contrast, the WD of V0515 at the surface is from approximately West-North-West (around 300 °). For both cases the wind speed (WS) at the surface is around 5 $\mathrm{m\,s^{-1}}$. The surface temperature of the sea ice is -6.15 °C and the temperature of the sea water is -1.7 °C (the melting point of standard ocean water). The model setup of this study is described in Section 3.4. In an additional set of simulations, which is described in the second part of this chapter (Section 5.3), the boundary layer (BL) structure near the surface is modified. The cloud layer is coupled to the surface to study the influences of the different surface types on the mixed-phase cloud.

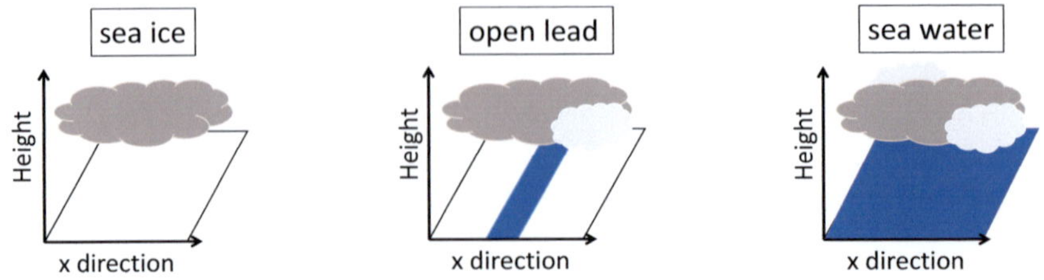

Figure 5.1: Sketch of the three different surface types of the model domain. The sea ice surface (white), the open lead surface, and the sea water surface (from left to right). The water area (blue) of the open lead surface takes 12.5 % of the whole surface area.

5.1 Influence of different surface types on the simulation of V0429

The mixed-phase cloud of V0429 is a low-level cloud with a cloud top at around 590 m (Chapter 4). The lowest atmospheric temperature of the initial profile is -5.8 °C , thus close to the sea ice surface temperature being defined as -6.15 °C (after Ovchinnikov et al. (2014)). Nevertheless, the sea ice is colder leading to a slightly negative and downward sensible heat flux at the surface (Figure 5.2a). The sea ice surface is warming over time because the sea water below the ice is warmer than the sea ice. Thus, the sensible heat flux is not increasing further with time. The latent heat flux is very small because of weak temperature differences between the surface and the atmosphere. The sensible heat flux and the latent heat flux over the sea water surface are both higher compared to the other two simulations (Figure 5.2c). The latent heat flux is constant around 11 W m^{-2}. In contrast, the sensible heat flux is decreasing with time because the sea water is warming the atmospheric layers near the surface, which is seen in the potential temperature (θ) profiles (Figure 5.3e). The latent heat flux and the sensible heat flux are clearly higher over the open lead area than over the sea ice surface (Figure 5.2d, e). Both fluxes over the open lead area reach values, which are similar to the fluxes over the sea water surface. Adding 12.5 % (approximately 1.5 W m^{-2}) of the mean latent heat flux over the open lead to 87.5 % (approximately 1.1 W m^{-2}) of the mean latent heat flux over the sea water lead almost to the latent heat flux over the

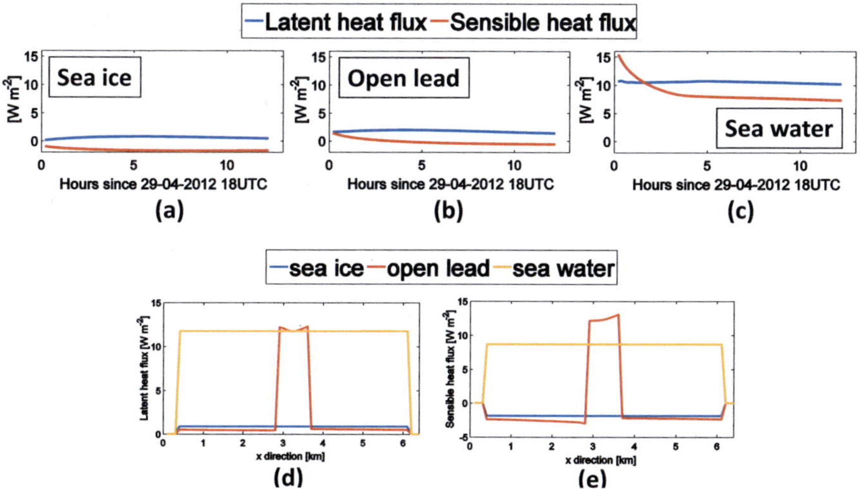

Figure 5.2: Domain average of the latent heat flux (blue) and the sensible heat flux (red) of the three decoupled simulations V0429 (sea ice surface **(a)**, open lead surface **(b)**, and sea water surface **(c)**). An upward flux is defined as positive. Average over the y-direction of the latent heat flux **(d)** and the sensible heat flux **(e)** after 6 h of simulation for all three surface types (colors).

open lead (Figure 5.2a-c). The same can be done for the sensible heat flux, where 12.5 % of the sea water surface sensible heat flux is around 1 W m^{-2} and 87.5 % of the sea ice simulation is around -1.3 W m^{-2}. By adding these values together, it is a slightly higher value than the simulated open lead sensible heat flux of around -0.5 W m^{-2}. Nevertheless, by combining the latent heat flux and the surface heat flux over the sea ice and over the sea water surfaces give a rough estimate of the surface fluxes over the open lead surface. In brief, the mean latent heat flux of the open lead surface is small, too. The mean sensible heat flux is slightly positive at the beginning of the simulation showing the influence of the warmer sea water temperature compared to the sea ice temperature (Figure 5.2b).

The differences in specific humidity (QV) profiles show only a small difference between the simulations over sea ice and over an open lead surface ($\Delta QV_{\text{openlead-seaice}}$) (Figure 5.4a). QV of the open lead simulation is increasing over time near the surface and the influence of the open lead in the surface on the near surface atmospheric levels is increasing. The simulations with the sea water surface show

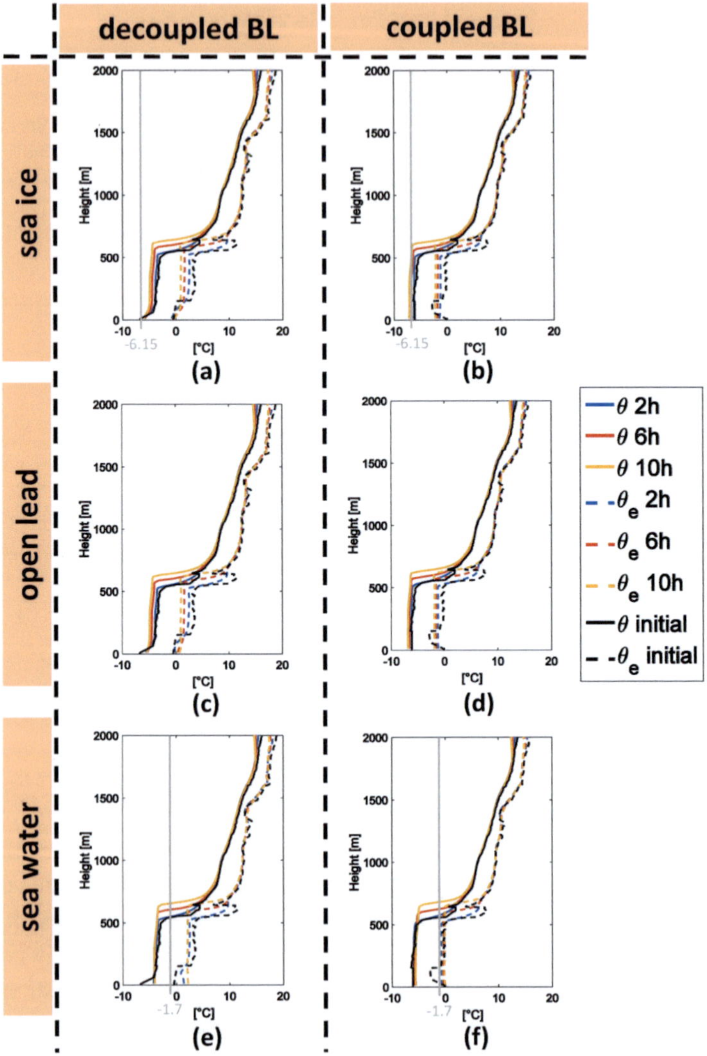

Figure 5.3: θ profiles (**(a)-(f)**, solid lines) and θ_e profiles (**(a)-(f)**, dashed lines) for the V0429 case. The surface types are sea ice (**(a)** and **(b)**), open lead (**(c)** and **(d)**), and sea water (**(e)** and **(f)**). The decoupled BL simulations are shown on the left side ((a), (c), (e)) and the coupled BL simulations (Section 5.3) are shown on the right side (**(b)**, **(d)**, **(f)**). The grey line marks the sea ice or sea water temperature.

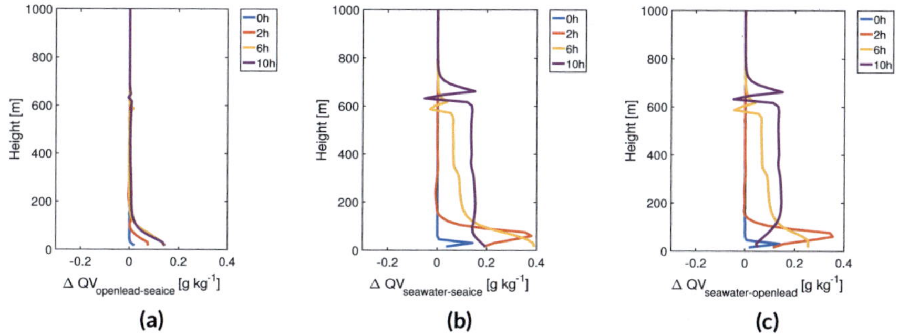

Figure 5.4: Domain average of ΔQV at four different time steps (after 0h: blue, 2h: red, 6h: yellow, and 10h: purple) for the differences in sea ice and open lead **(a)**, sea ice and sea water **(b)**, and open lead and sea water **(c)** for the decoupled simulations of V0429.

a stronger increase of QV near the surface by almost doubling ΔQV compared to $\Delta QV_{\text{openlead-seaice}}$ (Figure 5.4b and Figure 5.4c). It is clearly seen that QV is transported towards the cloud layer with time, which indicates a mixing of the surface moisture within the BL. The θ profiles and θ_e profiles of the three different simulations show that the decoupling of the BL from the surface remains in the sea ice simulations, becomes smaller in the open lead simulation, and the BL becomes coupled to the surface in the sea water simulations after two hours (Figure 5.3a, c, and e). The BL height increases as well as the inversion height and the BL becomes more mixed than initially in all three simulations (Figure 5.3a, c, and e).

Time series of the cloud liquid water path (LWP) of the three different simulations show almost no differences between the sea ice and open lead simulation (Figure 5.5, solid lines). The LWPs are similar at beginning of the simulations until around nine hours, then the open lead simulation performs a slightly higher LWP. After around three hours, the LWP of the sea water simulation increases to $50\,\text{g}\,\text{m}^{-2}$ and remains around this value. It can be summarized that the sea water surface influences the liquid in the cloud and leads to an increase in LWP of approximately 20 %.

The ice water path (IWP) are low and are around $1\,\text{g}\,\text{kg}^{-1}$ after three hours and $3\,\text{g}\,\text{kg}^{-1}$ at the end of the simulations for all three simulations (Figure 5.6, solid lines). During the simulations, more and more ice is produced and the increase of IWP is similar in the sea ice and open lead simulations. Contrary to this, the increase of IWP is faster in the sea water simulation within the first six hours. The condensation efficiency (CE) is calculated to show how much of the QV in the atmosphere is consumed to produce liquid and ice inside the cloud. Therefore, the ratio of LWP, IWP and water vapor path (WVP) is calculated in the BL up to the height of the cloud top (Equation (5.1)).

$$\mathrm{CE} = \frac{\mathrm{LWP} + \mathrm{IWP}}{\mathrm{WVP}} \tag{5.1}$$

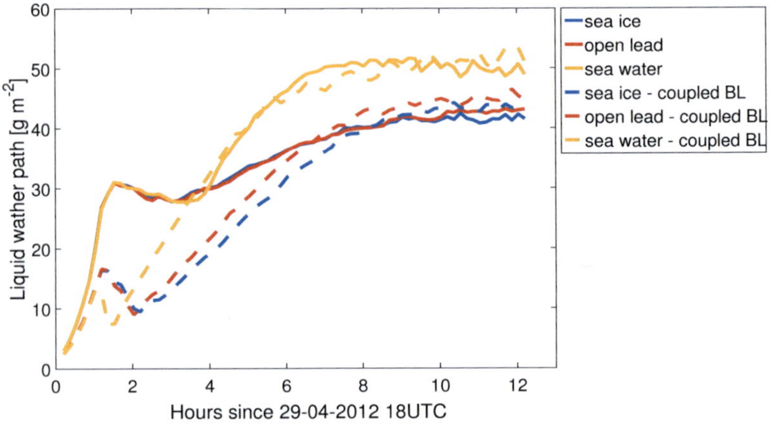

Figure 5.5: LWP time series of the simulations of V0429 with the 3 different surface types (colors) and the coupled BL (dashed lines) and decoupled BL (solid lines).

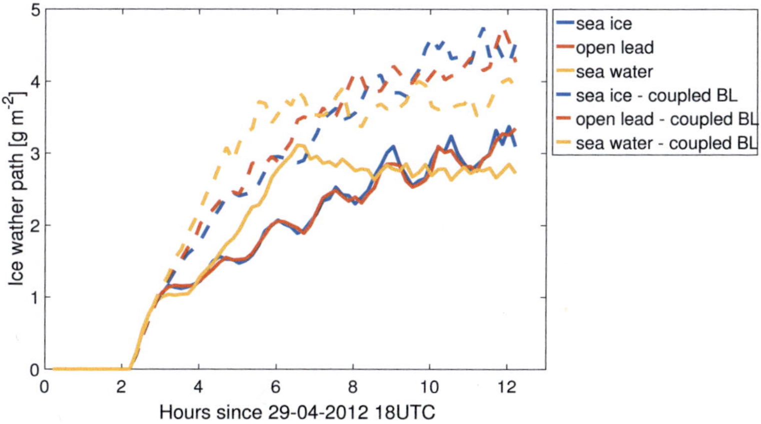

Figure 5.6: IWP time series of the simulations of V0429 with the 3 different surface types (colors) and the coupled BL (dashed lines) and decoupled BL (solid lines).

The time series of the simulations show a first CE maximum after around one hour of simulation and a decrease towards two hours (Figure 5.7, solid lines). After two hours, the ice processes are turned on (Section 3.4), thus the CE is increasing again because ice is produced. The strongest increase is seen in the sea water simulation, where the second maximum around 6 hours reaches the value of the first maximum. The other two time series (sea ice and open lead) show a small increase after two hours to only approximately 73 % of the first maximum of CE. Thus, the increase of moisture in the BL due to the increased surface fluxes over sea water surface have an impact on the cloud layer.

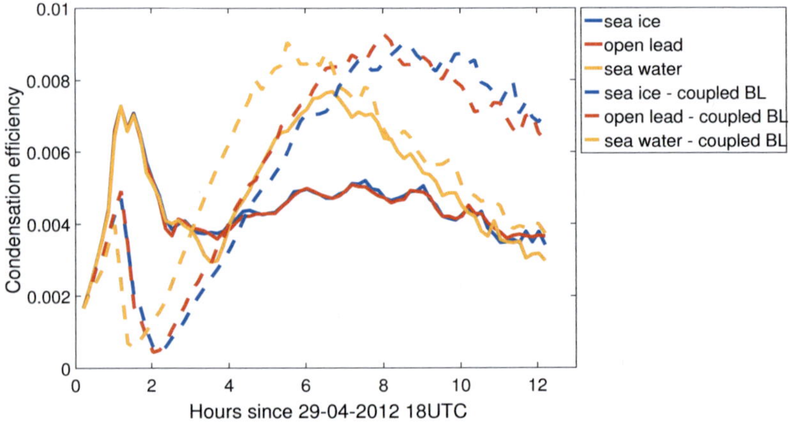

Figure 5.7: The condensation efficiency (Equation (5.1)) of the simulations of V0429 with the 3 different surface types (colors) and the coupled BL (dashed lines) and decoupled BL (solid lines).

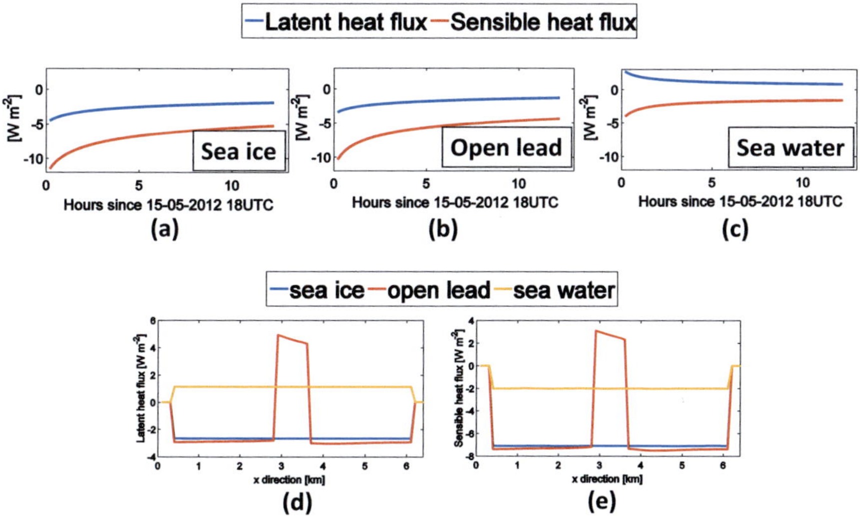

Figure 5.8: Domain average of the latent heat flux (blue) and the sensible heat flux (red) of the three decoupled simulations V0515 (sea ice surface **(a)**, open lead surface **(b)**, and sea water surface **(c)**). An upward flux is defined as positive. Average over the y-direction of the latent heat flux **(d)** and the sensible heat flux **(e)** after 6 h of simulation for all three surface types (colors).

5.2 Influence of different surface types on the simulation of V0515

The mixed-phase cloud of V0515 is a low-level cloud with a cloud top at around 990 m (Chapter 4). The lowest atmospheric temperature of the initial profile is -1.15 °C, thus 5 °C warmer than the sea ice temperature. Compared to the cloud in V0429, the cloud top is higher in V0515 and the temperature in the lowest atmospheric level is around 4.7 °C warmer than in V0429. Hence, the sensible heat flux is negative, especially over the sea ice and open lead surface (Figure 5.8). The sensible heat flux over the sea water surface is still negative but around 50 % smaller than the sensible heat fluxes of the two other simulations because the temperature difference between the surface and the atmosphere is smaller than over the other surfaces (Figure 5.8c). Over the sea water surface, the difference in moisture of the surface and of the atmosphere is small and leads to a positive and small latent heat flux (Figure 5.8c, d). The warmer atmospheric layers can store more moisture and thus the latent heat flux is directed downward over the sea ice and open lead surfaces. Nevertheless, a strong increase of latent heat flux and sensible heat flux is seen in the cross section in x-direction (Figure 5.8d, e). The latent heat flux of the open lead simulations can be almost calculated from 12.5 % of the sea water surface fluxes and 87.5 % of the sea ice surface fluxes (Figure 5.8b).

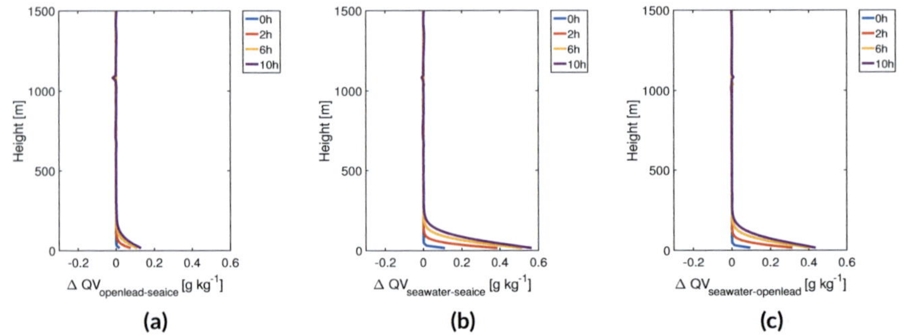

Figure 5.9: Domain average of ΔQV at four different time steps (after 0h: blue, 2h: red, 6h: yellow, and 10h: purple) for the differences in sea ice and open lead **(a)**, sea ice and sea water **(b)**, and open lead and sea water **(c)** for the decoupled simulations of V0515.

The ΔQV of the three different surface type simulations show that QV is highest near the surface (Figure 5.9). By comparing $\Delta QV_{\text{seawater-seaice}}$ and $\Delta QV_{\text{seawater-openlead}}$, which imply the sea water surface simulation (Figure 5.9b and Figure 5.9c), the ΔQV is more than twice as high as $\Delta QV_{\text{openlead-seaice}}$. Thus, the influence of the sea water surface is only seen in the near-surface atmospheric layers and not in the cloud layer. The moisture remains near the surface because of the small inversion near the surface in all three simulations (Figure 5.10a, c, and e). In general, the surface fluxes are small and they are not strong enough to erode the inversion near the surface. Hence, the provided moisture of the surface cannot reach the cloud layer.

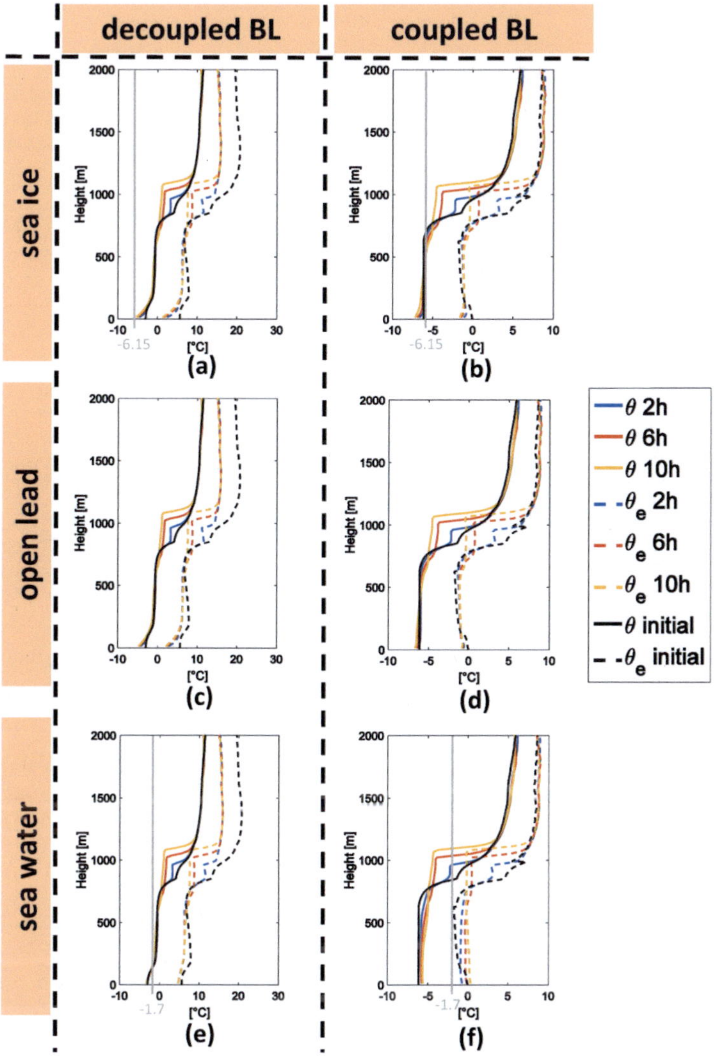

Figure 5.10: θ profiles ((**a**)-(**f**), solid lines) and θ_e profiles ((**a**)-(**f**), dashed lines) for the V0515 case. The surface types are sea ice ((**a**) and (**b**)), open lead ((**c**) and (**d**)), and sea water ((**e**) and (**f**)). The decoupled BL simulations are shown on the left side ((**a**), (**c**), (**e**)) and the coupled BL simulations (Section 5.3) are shown on the right side ((**b**), (**d**), (**f**)). The grey line marks the sea ice or sea water temperature.

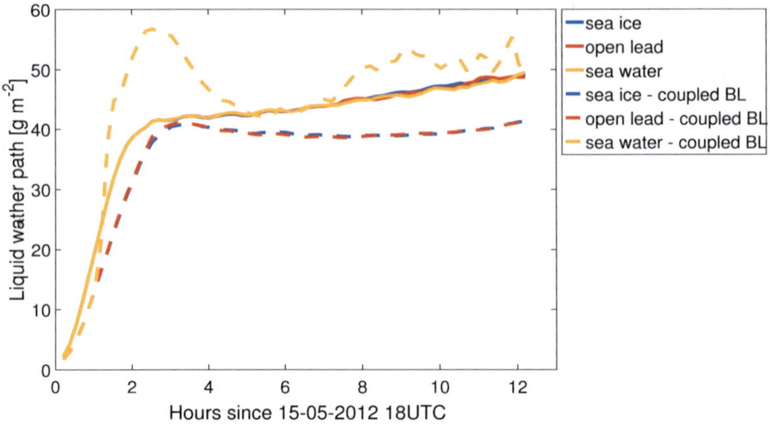

Figure 5.11: LWP time series of the simulations of V0515 with the 3 different surface types (colors) and the coupled BL (dashed lines) and decoupled BL (solid lines).

Additionally, the time series of the LWP show almost no differences between the three simulations, which indicates that no moisture from the surface reaches the cloud layer and consequently the cloud layer is decoupled from the surface and remains so for the entire simulation (Figure 5.11, solid lines).
Furthermore, almost no differences are seen in the time series of the IWP because of no differences in the LWP time series (Figure 5.12, solid lines). Therefore, the CEs of all three simulations are similar (Figure 5.13, solid lines).

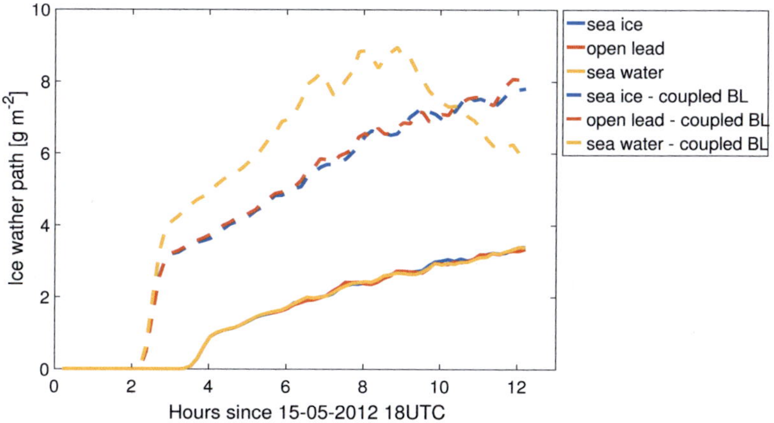

Figure 5.12: IWP time series of the simulations of V0515 with the 3 different surface types (colors) and the coupled BL (dashed lines) and decoupled BL (solid lines).

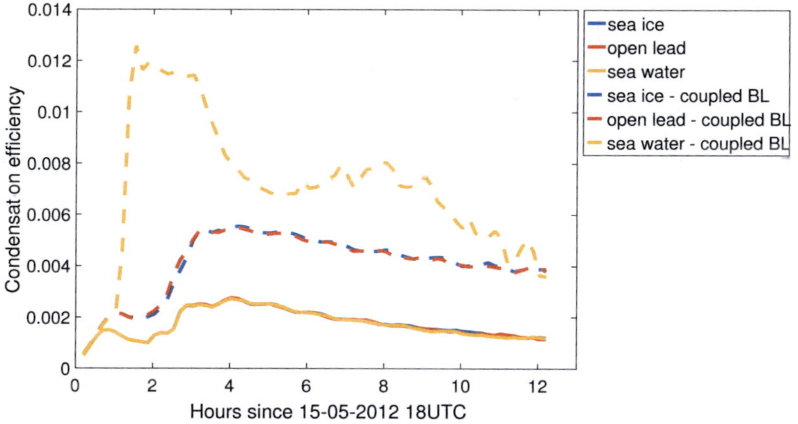

Figure 5.13: condensation efficiency of the simulations of V0515 with the 3 different surface types (colors) and the coupled BL (dashed lines) and decoupled BL (solid lines).

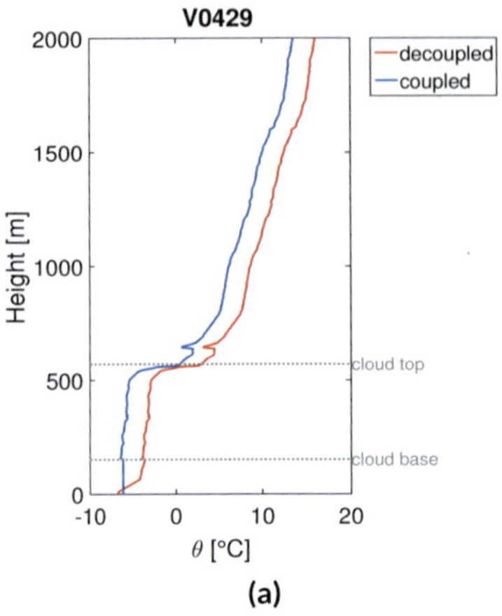

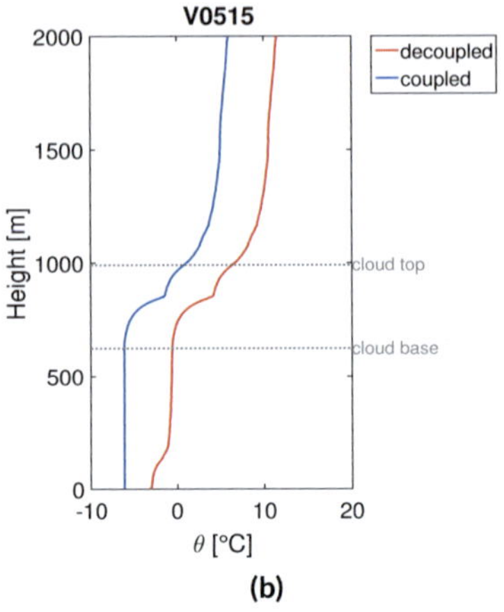

Figure 5.14: θ profiles of the decoupled (red) and coupled (blue) BL structures for V0429 **(a)** and V0515 **(b)**. Grey dotted lines mark the cloud top and cloud base.

5.3 Coupling the cloud layer to the surface

The previous section showed that the decoupling of the cloud layer from the surface prevents the moisture of the surface from mixing with the atmospheric levels above and with the cloud layer most of the time. To study the effect of a cloud which is coupled to the surface, the θ profiles of the two different VERDI simulations have been modified (Figure 5.14). For this, the first atmospheric level is set to the temperature of the sea ice surface (-6.15 °C) and a mixing layer is built towards the cloud base by modifying the θ profile. The θ at the first atmospheric level is held constant up to the cloud base height. Then, from the cloud base temperature upward, the whole θ profile is shifted by the difference between the decoupled cloud base temperature and the coupled cloud base temperature. The initial profiles of temperature and total specific humidity (water vapour and water) ($q_{\mathrm{tot_l}}$) are adjusted to these modified θ profiles such that the relative humidity (RH) stays the same as in the decoupled θ profile. With these adjusted profiles, three new simulations (with sea ice, open lead, and sea water surfaces) for each VERDI case are performed.

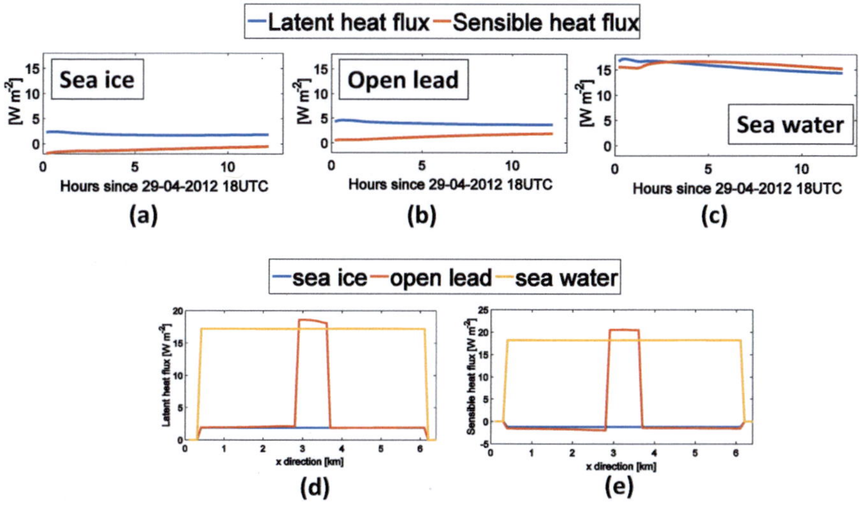

Figure 5.15: Domain average of the latent heat flux (blue) and the sensible heat flux (red) of the three coupled simulations V0429 (sea ice surface **(a)**, open lead surface **(b)**, and sea water surface **(c)**). An upward flux is defined as positive. Average over the y-direction of the latent heat flux **(d)** and the sensible heat flux **(e)** after 6 h of simulation for all three surface types (colors).

5.3.1 Influence of the coupled BL on the simulations of V0429

Due to the coupling of the BL to the surface, the BL is well-mixed from the surface level to the cloud base height. The surface fluxes of the coupled sea ice and open lead simulations are around twice that high as the decoupled simulations over these two surfaces (Figure 5.15a, b). The coupled sea water surfaces fluxes are positive from the beginning of the simulation and both, the latent heat flux and the sensible heat flux, are around 15 W m^{-2} (Figure 5.15c). The cross section in x-direction show that the latent heat flux and sensible heat flux are increased over the open lead area (Figure 5.15d, e). They are higher at this place than the surface fluxes of the sea water simulation.

For all three simulations, this mixing state remains over the whole simulation time (Figure 5.3). Due to the shifting of the initial θ profile to colder temperatures, the whole BL is colder compared to the decoupled simulations. The θ_e profiles reveal a shallow unstable layer with a inversion at around 200 m over the surface in the initial profile (Figure 5.3b, d, f). This inversion dissolves within two hours and evolves into a mixed layer. The BL over the sea ice becomes colder over time than in the other two simulations (Figure 5.3b). Over the sea water surface, the θ profiles are slightly warming with time (Figure 5.3f). Comparing the BL evolution of the decoupled simulations with the coupled simulations, the main differences are that the BL of the decoupled simulations needs time to evolve a mixed layer and the cloud layer is decoupled from the surface. In contrast to that, the simulations with a coupled BL have neutral θ profiles or slightly unstable θ_e profiles.

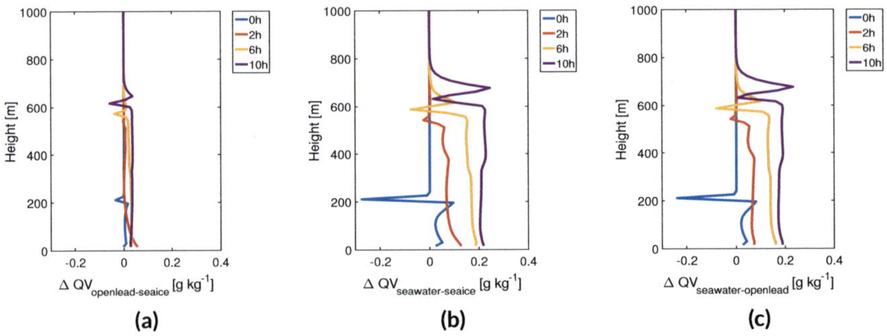

Figure 5.16: Domain average of ΔQV at four different time steps (after 0h: blue, 2h: red, 6h: yellow, and 10h: purple) for the differences in sea ice and open lead **(a)**, sea ice and sea water **(b)**, and open lead and sea water **(c)** for the coupled simulations of V0429.

The ΔQV profiles in all three simulations show the effect of the coupled BL because the QV is transported to the cloud layer and ΔQV is increasing with time, too (Figure 5.16). The sea water surface simulation shows that a considerable amount of QV is transported to the cloud layer. Around a quarter of QV of the sea water surface simulation is transported from the open lead surface to the cloud layer (Figure 5.16c). This is more than in the decoupled BL simulations, even the temperatures in the BL are colder. In general, ΔQV is smaller than in the decoupled BL simulations because of the colder temperatures.
The time series of LWP show that the sea water surface simulation has the highest LWP (Figure 5.5, dashed lines). The other two simulations show the same evolution of the LWP, but the LWP of the open lead simulation is around 2-3 g m^{-2}

higher than the LWP of the sea ice simulation. The difference to the decoupled simulations is small: Due to the colder temperatures, the atmosphere can store less QV than with warmer temperatures.
The IWPs of all three coupled simulations show a similar increase with time and all have higher IWPs than the decoupled simulations (Figure 5.6). This is due to the colder temperatures, thus more ice can form.
The CE gives an overview of how much condensate is produced during the coupled BL simulations (Figure 5.7). All coupled simulations produce more condensate than the decoupled simulations. In the first six hours, the sea water and open lead simulation have a higher CE than the sea ice simulation. Then, the CE of the sea water simulation decreases towards the CE of the decoupled simulations. The CE of the open lead simulations is similar to the CE of the sea ice simulations after six hours.

5.3.2 Influence of the coupled BL on the simulations of V0515

The initial θ in the lowest atmospheric level (around 15 m) has to be cooled by 5 °C to -6.15 °C being the temperature of the sea ice surface (Figure 5.14b). Thus, the BL temperatures differ from the decoupled simulations more than in the V0429 simulations, where the initial θ has to be cooled by approximately 2.5 °C. The surface fluxes of the coupled open lead and sea ice simulations are around 2-3 W m^{-2} higher than the surface fluxes of the decoupled simulations (Figure 5.17a, b). Compared to the decoupled sea water simulation the surface fluxes are both positive now and between 8-12 W m^{-2} (Figure 5.17c). The latent heat flux is twice as high over the open lead area compared to the decoupled simulation (Figure 5.17d). Furthermore, the sensible heat flux is around a third higher than the decoupled simulation (Figure 5.17e).
The initial neutrally stratified BL of the coupled V0515 simulations remains neutral until the end of the simulation only for the sea water surface (Figure 5.10f). The θ profiles show that the BL of the sea ice and open lead simulations is decoupling again from the surface (Figure 5.10b and Figure 5.10d). The θ_e profiles of all three simulations are initially slightly unstable near the surface and over time, the θ_e profile of the sea ice simulation becomes decoupled from the surface.

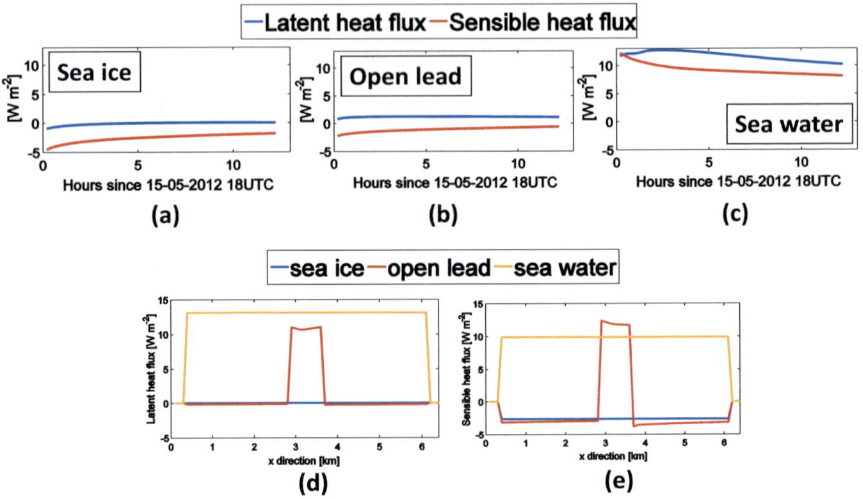

Figure 5.17: Domain average of the latent heat flux (blue) and the sensible heat flux (red) of the three coupled simulations V0515 (sea ice surface **(a)**, open lead surface **(b)**, and sea water surface **(c)**). An upward flux is defined as positive. Average over the y-direction of the latent heat flux **(d)** and the sensible heat flux **(e)** after 6 h of simulation for all three surface types (colors).

Furthermore, the θ_e profiles of the open lead and sea water simulation remain slightly unstable until the end of the simulation (Figure 5.10b and Figure 5.10d). Hence, the major difference between the coupled and decoupled θ and θ_e profiles is seen in the open lead and sea water simulations.

The ΔQV profiles show that QV, mainly from the sea water surface simulation, reaches the cloud base height (Figure 5.18). Surprisingly, more QV is seen at the surface of the sea ice and open lead than over the sea water surface (Figure 5.18b and Figure 5.18c). Due to the small latent heat flux over these two surfaces the initial QV remains near the surface (Section 4.3, Figure 4.9). This lasts only for around two hours then the BL becomes decoupled and QV is higher over the sea water simulations (Figure 5.10b, Figure 5.18b). More QV over the sea water and open lead surface is seen after two hours, then the QV reaches the cloud base. Conversely, the $\Delta QV_{\text{openlead-seaice}}$ profiles are all small at the surface and do not reach the cloud layer, instead the moisture remains near the surface and is higher over the open lead surface (Figure 5.18a). In general, QV is smaller than in the decoupled simulations.

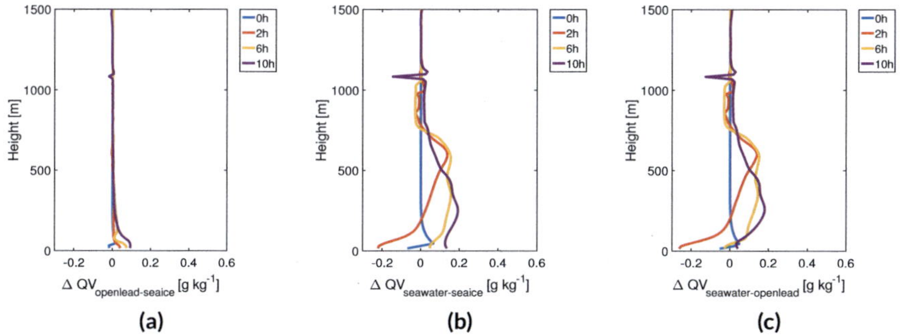

Figure 5.18: Domain average of ΔQV at four different time steps (after 0h: blue, 2h: red, 6h: yellow, and 10h: purple) for the differences in sea ice and open lead **(a)**, sea ice and sea water **(b)**, and open lead and sea water **(c)** for the coupled simulations of V0515.

The time series of the coupled simulations clearly show the highest LWP in the sea water simulation (Figure 5.11). The LWP time series of the sea water simulation increases fast at the beginning and reaches higher values than the decoupled simulations. After two hours when the ice processes are turned on (Section 3.4), the LWP decreases towards the values of the decoupled simulations. The LWP of the open lead and sea ice simulations are slightly smaller than the decoupled simulations.

A similar time series evolution is seen for the IWP of the coupled simulations (Figure 5.12). The IWP increase in the sea water simulation is faster than in the other two simulations and decreases towards the end of the simulation. In contrast, the IWP of the sea ice and open lead simulations are increasing during the whole simulation time whereas the IWP of the sea water simulation is not. All coupled simulations have higher IWP than the decoupled simulations because of the colder temperatures in the BL.

The colder temperatures influence the amount of IWP and LWP in V0515 more than in V0429. Thus, it is necessary to look at the CE time series (Figure 5.13). All coupled simulations have higher values of CE from the beginning of the simulations than the decoupled simulations. The increase of CE is highest in the sea water simulations and decreases after two hours when the ice particles start to grow. Compared to that, the CE of the sea ice and open lead simulations are increasing after two hours. Both, the sea ice and open lead CE time series are similar during the whole simulation time.

5.4 Discussion and Conclusions

In this sensitivity study, the influence of different surface types on Arctic mixed-phase clouds is analyzed. Therefore, two clouds (V0429 and V0515) during the VERDI campaign are simulated. The different surfaces cause differences in the sensible and in the latent heat fluxes. Especially, the domain average of the surface fluxes over the sea ice and open lead surface are small compared to the fluxes over the sea water surface. Due to the different BL temperatures in the two simulations, the mean surface fluxes can even be negative because the air temperature is warmer than the surface temperature. In all simulations the surface fluxes over the open lead area are increased and reach values similar to the surface fluxes over the sea water.

The BL structure near the surface shows the effect of the surface on the cloud layer. If the BL is decoupled from the surface, the moisture remains at the surface and cannot influence the cloud layer. Only if the BL is coupled to the surface during the simulation, QV can reach higher atmospheric layers and finally the cloud layer. This occurs in the V0429 simulations over the sea water surface. The surface fluxes influence the stability of the BL and allow low-level moisture to reach the cloud layer. In contrast, the BL of V0515 remains decoupled from the surface over all three surface types because the inversion near the surface is around 100 m higher than in the V0429 simulation. Hence, the moisture does not reach the cloud layer because the surface fluxes are small or negative which affects the BL structure at the surface. Thus, no influence of the surface is seen in the cloud layer.

The impact of the surface on the cloud properties can be analyzed with the CE. The decoupled simulations are all in the same range and show a similar evolution over time. Except for the V0429 simulation over sea water, where the BL structure changes with time and the BL becomes coupled to the surface. Hence, the influence of the sea water surface becomes relevant. More QV can reach the cloud layer leading to a higher CE.

All coupled simulations have higher values of CE than the CE of the decoupled simulations after four hours or earlier. The effect of the mixed BL in combination with different surfaces is seen over all surfaces but most strongly over the sea water surface. Interestingly, only small differences in CE are seen in the open lead and sea ice simulations. Thus, the influence of the open lead on the cloud properties is small. Only the sea water surface simulation shows a considerable effect, which is more pronounced in the V0515 simulations because the BL evolves from a neutral to a slightly unstable layer. The CE of the V0429 coupled and decoupled sea water simulations become similiar towards the end of the simulation because the decoupled simulation becomes coupled to the surface after two hours.

To come back to the research question raised in Chapter 1, the results presented here suggest to draw the following conclusions:

Do different surface types influence the cloud properties of an Arctic mixed-phase cloud?

- The BL structure at the surface as well as the temperatures above the surface layer are important and influence the intensity of the surface fluxes.
- More QV can reach the cloud layer when the BL is mixed from the cloud base height towards the surface.
- The sea water surface leads to a coupling of the BL in certain scenarios (if the surface inversion is near the surface) even if the BL was decoupled before.

CHAPTER 6

The dissipation of an Arctic mixed-phase cloud

Arctic mixed-phase clouds are long persistent and thus it is important to investigate the processes which lead to eventual dissipation of such a cloud. This chapter describes the simulations of an Arctic mixed-phase cloud during the Arctic Summer Cloud Ocean Study (ASCOS) campaign and studies the processes which lead to dissipation of the Arctic mixed-phase cloud observed on 31 August 2008. The results are published in Atmos. Chem. Phys. Discuss. (Loewe et al., 2016) and accepted in Atmos. Chem. Phys.; The following text and graphics are based on this publication.

6.1 The Arctic Summer Cloud Ocean Study (ASCOS)

The ASCOS campaign took place during summer 2008, and the entire expedition lasted more than one month in the central Arctic Ocean of the North Atlantic (Tjernström et al., 2012). Detailed boundary layer (BL) and cloud measurements were taken when the Swedish icebreaker *Oden* drifted with a multi-year ice floe for three weeks around 87° N (Tjernström et al., 2014). This study will focus on an episode towards the end of the ice drift, around 31 August 2008 (DoY 244).

A low-level stratiform cloud layer had been quasi-persistent for about one week but dissipated rapidly in the evening of DoY 244 (Sedlar et al., 2011; Mauritsen et al., 2011; Sotiropoulou et al., 2014). The period of this persistent stratocumuls layer (DoY 236–244) (Fig. 6.1) was dominated by a quasi-steady high pressure system, with passages of a few weak fronts (Tjernström et al., 2012). The surfaces temperature was around 2 °C, hence around the freezing point of salt water. The transition to autumn freeze-up happens towards the end of the period. The BL can be divided in 3 different layers (Tjernström et al., 2012). The first shallow surface-based layer was approximately 200 m deep and the turbulence was driven by wind shear. In the well-mixed layer from 800 to 1000 m, related to the cloud layer, buoyant overturning driven by longwave cooling generated the turbulence. Between these two layers a near-neutral stratified layer is located. The wind shear was on average 0 above 200 m.
Mixed-phase stratiform clouds often tend to be decoupled from surface layer turbulence by a statically stable layer. During ASCOS, low-level mixed-phased clouds were decoupled from the surface about 75 % of the time (Shupe et al., 2013; Sedlar and Shupe, 2014; Sotiropoulou et al., 2014). The cloud layer shown in Fig. 6.1 was decoupled during the 8 h period of interest (Shupe et al., 2013). Observations from the vertically pointing Doppler millimeter cloud radar (MMCR) shows the cloud top at around 1 km during the morning hours, with a thinning and lowering cloud top during the afternoon (Fig. 6.1). The laser ceilometer measured the cloud base at around 600 m to 700 m in the morning. During the day, the cloud base decreased towards the surface. With observations from the MMCR, a dual-channel microwave radiometer (MWR), a ceilometer, and radiosondes, the cloud type was classified as mixed-phase during the first half of the 31 August 2008. The cloud type classification follows the method by Shupe (2007). The retrieval of the liquid water path (LWP) from the MWR contains of an uncertainty of 25 g m-2 (Westwater et al., 2001), explaining the negative LWP observations in Fig. 2a. During DoY 244, the LWP increased from around 90 g m^{-2} to values over 300 g m^{-2} and varied considerably during the first half of the day. Finally, the LWP reached values around 50 g m^{-2} in the afternoon. The ice water path (IWP) is integrated from profiles of the ice water content (IWC), which are derived from MMCR reflectivity power-law relationships at vertical levels deemed to predominantly ice-phase by the cloud phase classifier (Shupe et al., 2005, 2006). The uncertainty in IWC retrieval, as large as a factor of two (Shupe et al., 2005,

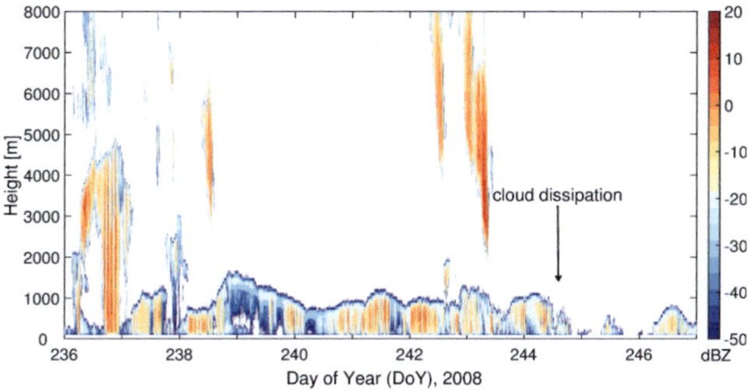

(a)

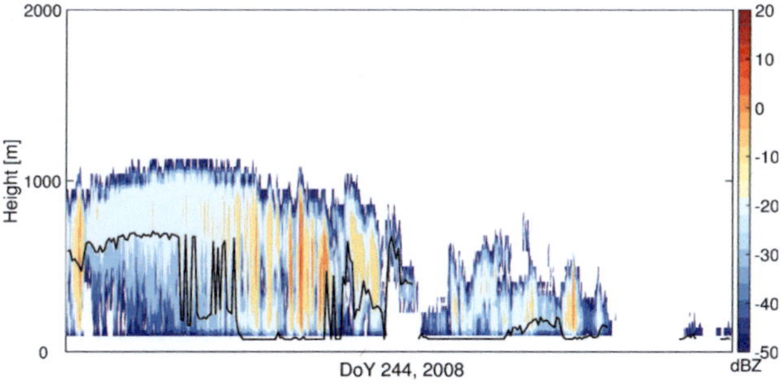

(b)

Figure 6.1: Radar reflectivity factor (colours, dBZ) contour time series for the period DoY 236 to 246 during 2008 **(a)** and for DoY 244 **(b)**. The ceilometer measurements of the cloud base height is shown with the black solid line in **(b)**.

2006) results from a combination of systematic and random errors. The IWP was in the range of 10 g m^{-2} in the morning and varied over a wide range until 12 UTC. After 12 UTC the IWP was around or even below 5 g m^{-2} (Fig. 6.2).
A cloud condensation nuclei (CCN) counter fed from an inlet on *Oden* approximately 20 m above the surface measured a mean CCN concentration of about 25 cm^{-3} at a supersaturation of 0.2 % during the time period of the ice drift (DoY 225–246) (Martin et al., 2011). During the evening of DoY 244, CCN concentrations at the surface dropped below 1 cm^{-3} around the time that the cloud began to dissipate (Mauritsen et al., 2011; Leck and Svensson, 2015). It is important to

understand that since the cloud layer was decoupled, at least initially, we do not know how representative these values are for the cloud layer.

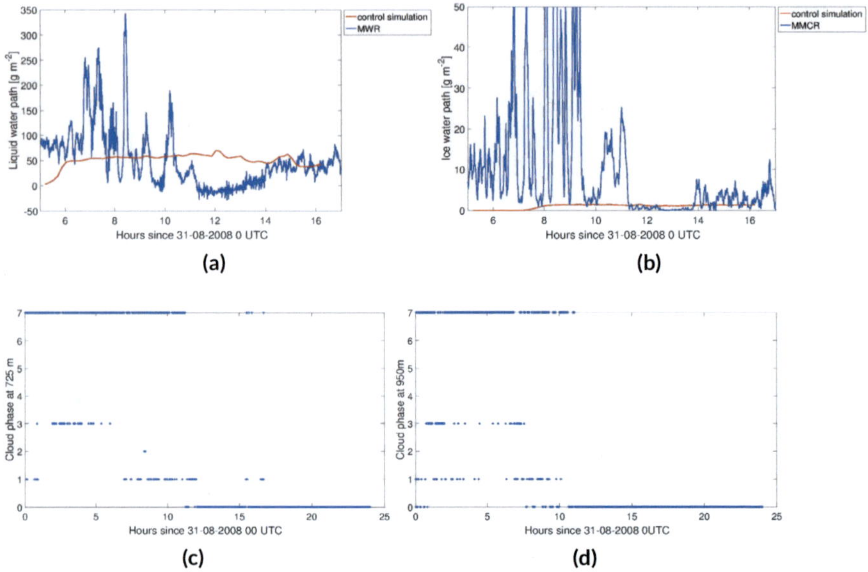

Figure 6.2: The observed and simualted LWP and RWP **(a)** and the IWP **(b)** is shown in blue and red, respectively. The simulated LWP includes cloud droplets and raindrops. In the bottom row the cloud phase classification is shown for two different heights, 725 m **(c)** and 950 m **(d)**, during the 31 August 2008. The classification number is described as follows: 0 - clear, 1 - ice, 2 - snow, 3 - liquid, 4 - drizzle, 5 - liquid cloud + drizzle, 6 - rain, 7 - mixed-phase, 8 - haze, 10 - uncertain.

6.2 Model setup of the sensitivity experiments

The model study was divided into one control simulation and three sets of sensitivity experiments (Tab. 6.1). The initial profiles were the same in all simulations except for the moisture profiles of sensitivity experiment SensMoist. A detailed description of the model setup is found in Section 3.4.

Table 6.1: Overview and short description of the different sensitivity simulations with specifications of the cloud droplet number concentration (CDNC), ice crystal number concentration (ICNC), and relative humidity (RH).

Simulation	Specifications
Control simulation	CDNC = 30 cm^{-3}, ICNC = 0.2 l^{-1}
Sensitivity experiment 1 (SensMoist)	dry-air advection below the cloud
	dry-air advection above the cloud with RH of 36 %
	dry-air advection above the cloud with RH of 20 %
	dry-air advection above the cloud with RH of 10 %
Sensitivity experiment 2 (SensIce)	ICNC = 1 l^{-1}
	ICNC = 10 l^{-1}
Sensitivity experiment 3 (SensCDNC)	CDNC = 10 cm^{-3}
	CDNC = 2 cm^{-3}

6.2.1 Control simulation

The control simulation has a CDNC of 30 cm^{-3} and a fixed ICNC of 0.2 l^{-1}; these values are chosen based on the mean values during ASCOS reported in Section 6.1. Observations at the surface did not record any ice nuclei (IN) concentrations because they were below the detection limit of the instrument, which ranges between 0.1 l^{-1} and 2 l^{-1} (Z. Kanji, personal communication). However, the fact that clouds during ASCOS precipitated predominantly ice crystals implies that IN must have been present (Shupe et al., 2013). It is possible that advection with or without entrainment of IN at cloud top rather than surface sources provided IN for the observed cloud. An earlier field campaign with *Oden* during September 1991 measured a maximum ice-forming nuclei concentration of 0.25 l^{-1} at 88° N (Bigg, 1996). Guided by these findings, the ice crystal concentration in the model was set to be 0.2 l^{-1} in the control simulation.

6.2.2 Sensitivity experiments

The first sensitivity experiment (SensMoist) includes 4 simulations where the moisture profile is changed either below the cloud base (sub-cloud layer) or above the cloud top in order to mimic the effect of dry-air advection (Tab. 6.1). Below cloud, the moisture profile is linearly dried to resemble 99 % RH at cloud base decreasing to 85 % RH at the surface (Fig. 6.3, a), while keeping the temperature profile the same as in the control simulation. Above cloud top, a 450 m deep layer of the atmosphere is progressively dried in 3 different simulations corresponding to RH values of 36 %, 20 % and finally 10 % above and in contact with cloud top (Fig. 6.3, b-d).

In order to investigate the sensitivity of the modeled cloud to changes in ice crystal concentrations, the ice crystal concentration was increased to values well above the expected low values in the Arctic in the second set of sensitivity experiments (SensIce). Two simulations with ice crystal concentrations set to $1\,l^{-1}$ and $10\,l^{-1}$ were conducted.

The third sensitivity experiment (SensCDNC) considers the low CCN concentrations observed during the ASCOS field campaign. During DoY 244, CCN concentrations at the surface dropped below $1\,cm^{-3}$ (Mauritsen et al., 2011; Leck and Svensson, 2015). The CDNC was decreased to $2\,cm^{-3}$ and $10\,cm^{-3}$, respectively, in two simulations to investigate the impact of low CDNC on the mixed-phase cloud development.

6.3 Results of the sensitivity experiments

6.3.1 Control simulation

The initial θ-profile shows a neutral to stable BL (Fig. 6.4a). A small inversion is seen in both θ and θ_e profiles near 300 m; this is the decoupling inversion, separating turbulence driven by the cloud layer from surface-driven turbulence. After 4 h of simulation, the model θ_e-profiles display a well-mixed layer extending from the surface to the inversion base near 1 km; this is also where the cloud top is located (Fig. 6.5). The sensible heat flux and the latent heat flux are weak, because the surface is covered with ice (Fig. 6.6). The observed values of both of these fluxes are small, but positive (Sedlar et al., 2011). Thus, no strong influence of the surface on the cloud is expected. The simulated cloud top corresponds well with the cloud top seen by the MMCR at around 1 km (Fig. 6.1, b). The cloud base, measured with a laser ceilometer, is between 600 m and 700 m at the beginning of DoY 244. This altitude agrees well with the cloud base height of the simulated cloud layer, which is around 600 m (Fig. 6.5). The BL deepens over the next 8 h, causing the main inversion and the cloud top to rise by around 90 m (Fig. 6.5).

The θ profiles imply that the lower half of the BL transitions towards less stable and hence the decoupling inversion around 300 m disappears after 4 h (Fig. 6.4a). The boundary and cloud layers thus quickly become coupled in the simulations. This tendency to erode cloud decoupling is common in large eddy simulations (LES) simulations (Savre et al., 2014).

The maximum ice water content (IWC) in the control simulation is two orders of magnitude smaller than the liquid water content (LWC), around 0.0015 $g\,kg^{-1}$ (Fig. 6.5). After 3 h the mixed-phase cloud generates ice and begins precipitating ice crystals which fall through the sub-cloud layer and reach the surface (Fig. 6.5, red). At the same time a secondary cloud layer briefly forms at the decoupling inversion, likely associated with a moistening of the sub-cloud layer through ice crystal sublimation. Rain (rain water content (RWC)) precipitates out of the liquid layer after around 7 h but does not reach the surface due to evaporation and conversion to ice (Fig. 6.5, green). After 4 h, when ice formation is relatively constant, the control simulation develops a liquid cloudy layer that is persistent throughout the simulation with a thickness of approximately 200 m (Fig. 6.5, blue). The maximum LWC in the cloud is around 0.2 $g\,kg^{-1}$. Observations of the LWP show a more variable LWP in the morning than in the afternoon (Fig. 6.2). The simulated LWP is around 50 $g\,m^{-2}$ and most of the time in the range of the observed LWP, which has an error of 25 $g\,m^{-2}$. Because the simulated cloud is not dissipating in the control simulation, the simulated LWP remains in that range and is not decreasing during the day. The IWP of the control simulation seems to be at the lower end of the observed IWP range and reaches only around 2 $g\,m^{-2}$ after the ice processes are turned on.

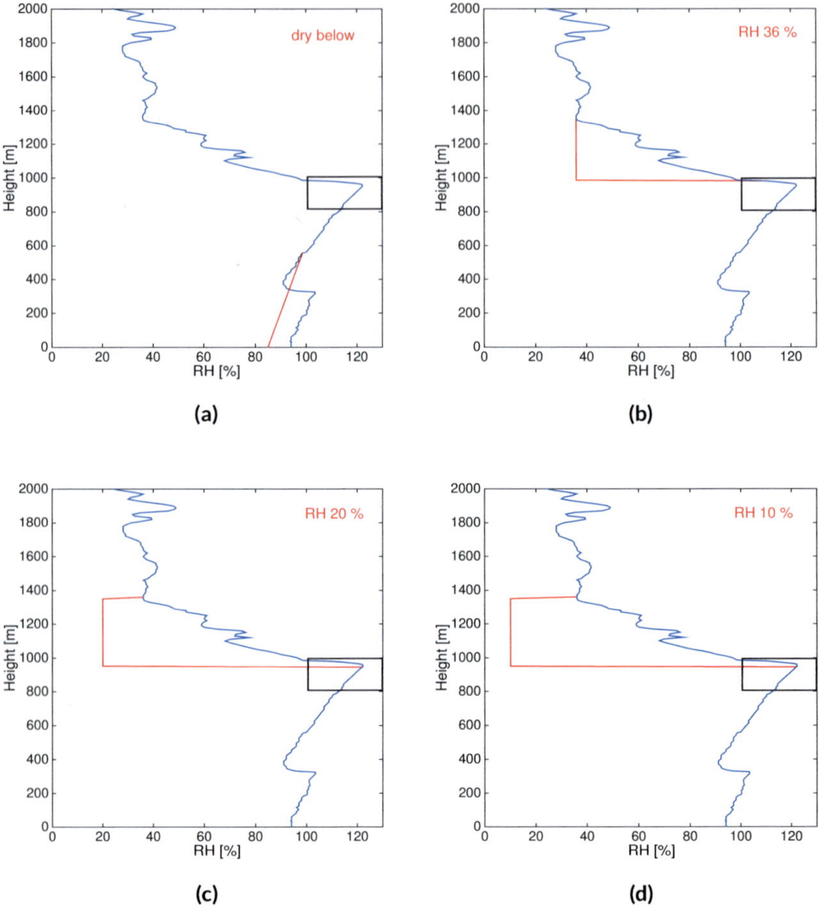

Figure 6.3: RH profiles for the sensitivity experiment SensMoist simulations. Modified parts of the RH profile are shown in red. The initial RH profile is in blue. **(a)** RH profile for the simulation with dry-air advection below the cloud. **(b)** shows the RH profile of the simulation with the dry-air advection above the cloud with the RH of 36 %, **(c)** and **(d)** show simualtions with the dry-air advection above the cloud with the RH of 20 % and 10 %, respectively. Black box marks the vertical extent of the liquid cloud layer of the control simulation.

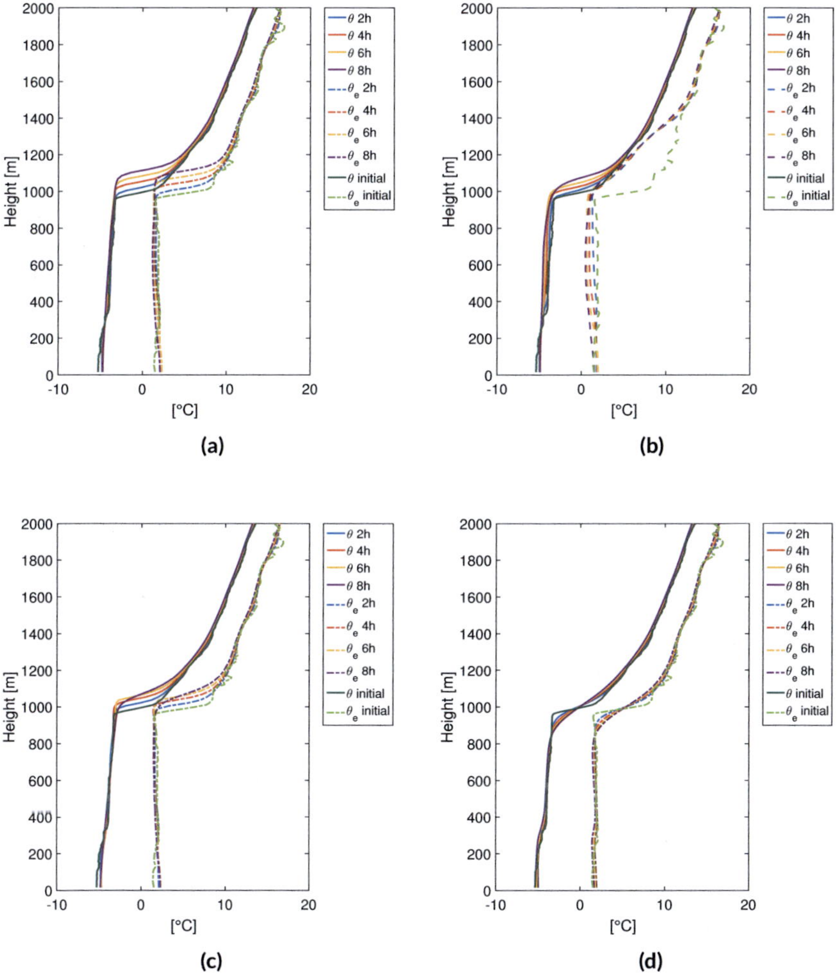

Figure 6.4: θ (left) and θ_e (right) profiles of the initial conditions (dark green and green), and domain averaged profiles after 2 h (blue), 4 h (orange), 6 h (yellow) and 8 h (purple) from the start of the control simulation **(a)**, the RH 10 % simulation **(b)**, the ICNC=10 l^{-1} simulation **(c)**, and the CDNC=2 cm^{-3} simulation **(d)**.

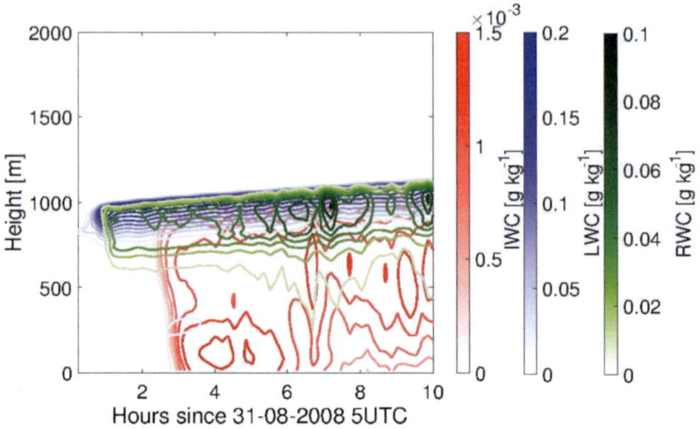

Figure 6.5: Domain averages of LWC (blue), IWC (red) and RWC (green) of the control simulation.

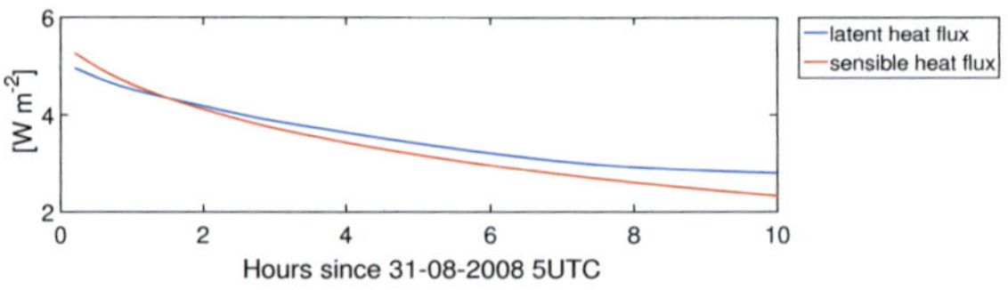

Figure 6.6: Domain averaged latent heat flux (blue) and sensible heat flux (red) of the control simulation.

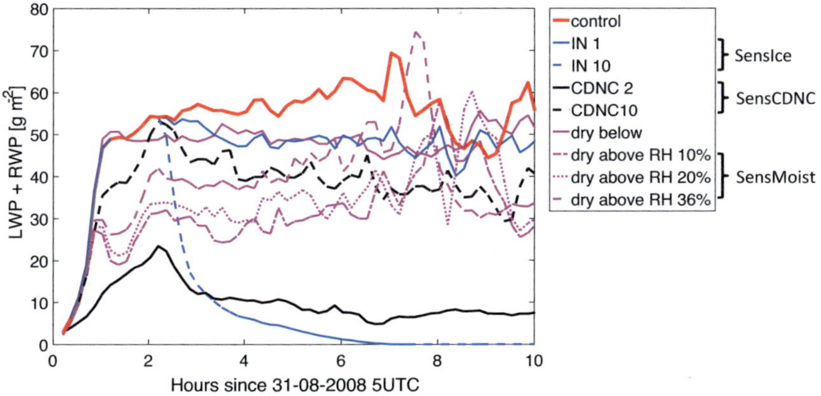

Figure 6.7: The domain averaged LWP and RWP for the control simulation (red), the simulation of dry-air advection below the cloud (pink solid), the simulations of dry-air advection above the cloud top with a RH of 36 % (pink dashed), a RH of 20 % (pink dotted) and a RH of 10 % (pink dash-dotted), the simulations with an ICNC of $1\,l^{-1}$ (blue solid) and $10\,l^{-1}$ (blue dashed), the simulation with a CDNC of $10\,cm^{-3}$ (black dashed) and a CDNC of $2\,cm^{-3}$ (black solid).

6.3.2 Sensitivity experiment - SensMoist

The availability of moisture above and below the cloud is an important ingredient for the persistence of an Arctic mixed-phase cloud. Fig. 6.7 shows the evolution of LWP in the SensMoist experiment (pink lines). Reducing the available moisture in the atmosphere below the cloud does not change the persistence of the cloud. Up to 8 h, the LWP of the simulation with reduced moisture below the cloud is slightly smaller (by approximately $8\,g\,m^{-2}$ compared to the control simulation after 4 h) than in the control simulation and at the end of the simulation, the LWP is almost the same (Fig. 6.7, pink solid line). The mean profiles of specific humidity (QV) after 5 h of simulation show that the difference in moisture is small near the surface between the different simulations indicating a strong mixing in the sub-cloud layer (Fig. 6.8). A strong difference in QV is only seen above the cloud top and between the control simulation and the two SensMoist simulations with reduced RH. This suggests that the supply of moisture from near the surface has only a limited influence on the cloud layer, resulting in a stable LWP around $50\,g\,m^{-2}$. Imposing a region of dry-air above the cloud has a larger influence on the cloud evolution. Drier air above the cloud layer leads to a decrease in LWP in all three simulations (Fig. 6.7, dashed pink lines). The reduction is strongest when RH above the cloud was reduced to 10 %. The LWC is reduced by almost a factor of 2 compared to the control simulation (Fig. 6.9). When the source of moisture from above is decreased, the BL and cloud layer become coupled between 2 h and 4 h, which is similar to the control simulation (Fig. 6.4b).

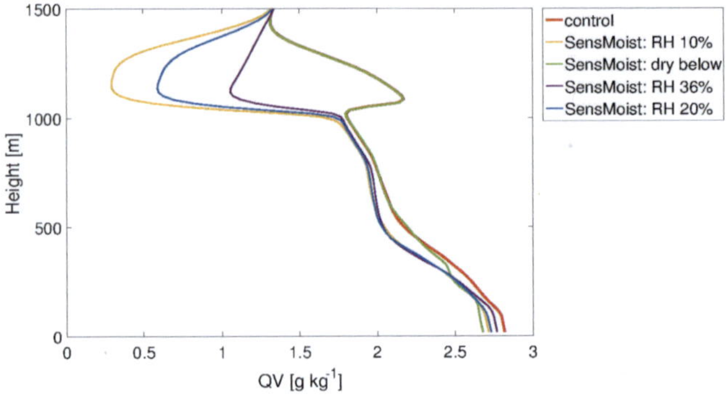

Figure 6.8: Domain averaged profiles of the water vapor (QV) are shown after 5 h of simulation for the control simulation (red) and the SensMoist simulations (RH 10 %: yellow, dry below: green, RH 36 %:purple, and RH 20 %: blue).

The θ_e profiles also show a clear weakening of the inversion after 2 h which is due to the thinner cloud layer and consequently decreased turbulence (Fig. 6.4b). Hence the BL cannot grow with time as it does in the control simulation (Fig. 6.5, 6.9). Following the reduction in LWC, IWC is also reduced relative to the control simulation; the mass of the liquid droplets is decreased and therefore ice crystals grow less rapidly. This causes the ice crystals to remain suspended in the atmosphere longer due to their reduced size and fall speed (6.9, red). These results examining the sensitivity of cloud to the moisture profile changes agree with the behavior of the Arctic mixed-phase cloud as reported in Solomon et al. (2013).

6.3.3 Sensitivity experiment - SensIce

In the simulation with an increased ice crystal concentration to $1\,\mathrm{l}^{-1}$, the cloud is still persistent over the simulation time, and IWC increases because of the large

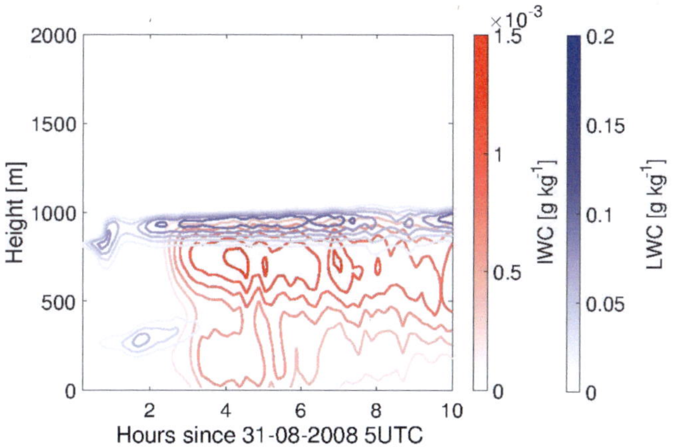

Figure 6.9: Domain averages of LWC (blue) and IWC (red) of the SensMoist simulation with the RH of 10 % above the cloud top.

number of ice crystals (Fig. 6.10). The impact on the liquid layer however is marginal. The LWP is almost constant at around 50 g m^{-2}, very similar to the LWP evolution simulated when RH is reduced in the sub-cloud layer (Fig. 6.7). Further increasing the ice crystal number to 10 l^{-1} leads to glaciation and finally to dissipation of the cloud after 6 h (Fig. 6.7, blue dashed line). In this simulation, the inversion near cloud top becomes weaker after 8 h and the weak stable layer near 300 m erodes more rapidly than in the control simulation (Fig. 6.4c).

6.3.4 Sensitivity experiment - SensCDNC

When CDNC is reduced relative to the reference value in the control simulation, the LWP time series shows a decrease to around 40 g m^{-2} with the CDNC 10 cm^{-3}, and to below 10 g m^{-2} for CDNC set to 2 cm^{-3} (Fig. 6.7, black lines). The reduction in CDNC also leads to a weakening of the inversion around 1 km, while the inversion near 300 m persists throughout the simulated duration, whereas it is eroded after roughly 4 h in the control simulation (Fig. 6.4d). The weakening of the main inversion is likely due to less radiative cooling at the cloud top, because of the optically thinner, less opaque liquid layer due to lower CDNC. This also decreases the cloud overturning circulation which in turn slightly strengthens the decoupling inversion. The cloud-top radiative cooling is reduced, and subsequently the cloud-driven circulation is unable to sufficiently penetrate the static stable layer near 300 m. With an optically thinner cloud above, the sub-cloud layer can cool more efficiently, and this promotes the formation of secondary, thin liquid layer in the vicinity of the lower temperature inversion near 300 m (Fig. 6.11). Rain forms after 2 h from initialization, through collision and coalescence processes. Rain from the main cloud layer can moisten the sub-cloud layer due to evaporation until the cloud layer at 1 km almost dissipates. This simulation leads to a very thin cloud with LWC values reaching $0.03\,\mathrm{g\,kg^{-1}}$ and maximum values of IWC of $0.0015\,\mathrm{g\,kg^{-1}}$ close to the surface. Ice crystals falling from the upper cloud layer pass through the lower liquid layer around 3 h simulation time, where they grow at the expense of cloud droplets, resulting in IWCs as large as the control simulation. This also causes the second, lower liquid cloud to become tenuous and briefly intermittent (Fig. 6.11).

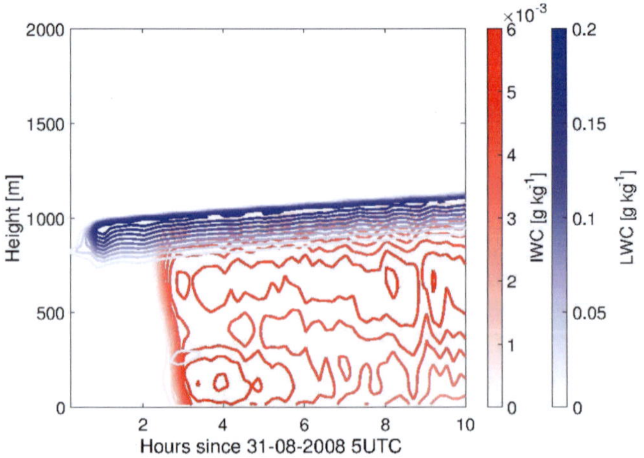

Figure 6.10: Domain averages of LWC (blue) and IWC (red) of the SensIce simulation with an ice crystal concentration value of $1\,\mathrm{l^{-1}}$.

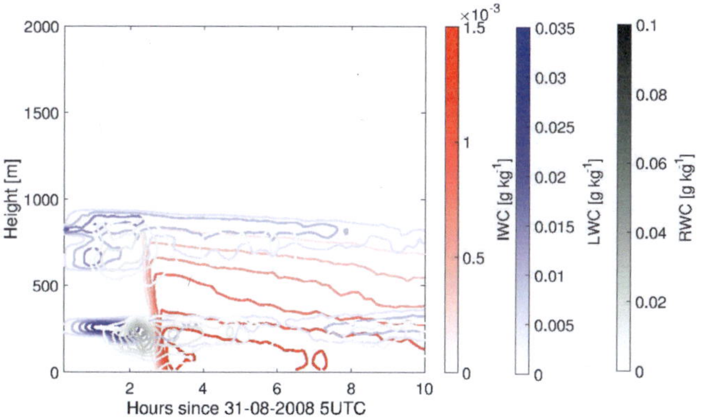

Figure 6.11: Domain averages of LWC (blue), IWC (red) and RWC (green) of the SensCDNC simulation with 2 cm^{-3} CDNC.

6.4 Discussion and Conclusions

Low aerosol concentrations are common in the high Arctic due to a lack of aerosol sources in this region in particular during summer (Bigg, 1996; Heintzenberg et al., 2006; Garrett et al., 2010; Heintzenberg and Leck, 2012). In persistent precipitating boundary layer clouds, the aerosol concentration can be further reduced through scavenging. Thus, changes in aerosol concentrations and consequently CDNCs may strongly influence the lifetime and development of an Arctic mixed-phase cloud. The current model study of an observed mixed-phase cloud during the ASCOS field campaign shows that a CDNC concentration of 10 cm^{-3} is sufficient to sustain the cloud while a CDNC of 2 cm^{-3} leads to cloud dissipation.

The results are in agreement with Mauritsen et al. (2011), who discussed a tenuous cloud regime in the Arctic characterized by low CCN number concentrations. Mauritsen et al. (2011) found that a CCN number concentration of 10 cm^{-3} marked the upper boundary for a transition regime below which cloud formation becomes limited. Observations during three previous campaigns in the high Arctic and during the ASCOS field campaign indicate a 25 to 30 % occurrence frequency of CCN concentrations below this value, i.e. within the so-called tenuous cloud regime (Mauritsen et al., 2011); during ASCOS the median CCN concentration was 20 to 30 cm^{-3} as measured by two independent CCN counters set to the same supersaturation; CCN concentrations were below $10\,cm^{-3}$ about 20 to 30 % of the time (Tjernström et al., 2014).

Using both a three-dimensional and a single-column model (SCM) version of the Met Office Unified Model (MetUM) numerical weather prediction model, and exploring an extended period of ASCOS observations, Birch et al. (2012) found that a constant CCN concentration of $10\,cm^{-3}$ instead of $100\,cm^{-3}$ gave a better general representation of low-level mixed-phase cloud properties. In a study of Arctic stratocumulus clouds and dynamic surface coupling, Sotiropoulou et al. (2014) used an indirect method to show that that the presence of optically thin clouds observed during ASCOS correlate with low CCN concentrations and persist for about 30 % of the time. The analysis by Mauritsen et al. (2011), Birch et al. (2012), and the findings in the present study indicate that a drop in aerosol and CCN number concentration to values below $10\,cm^{-3}$ may be an important reason for mixed-phase cloud dissipation in the high Arctic in summer. It is therefore important that models, in particular with interactive aerosol and cloud microphysics, can represent this type of low aerosol cloud regime, while many models assume constant droplet number or aerosol concentrations representative for mid-latitudes (Wesslén et al., 2014; Sotiropoulou et al., 2016).

While Birch et al. (2012) ran their simulations over several days, the COSMO simulations presented here are for 10 h and focused on the cloud development during that time. With the high horizontal and vertical resolution, the COSMO simulations focus on the cloud microphysics and cloud evolution over a shorter time compared to the MetUM SCM simulations of Birch et al. (2012). The idealized setup with periodic boundary conditions and the small domain limit the investigation area and hence focus only on parts of the Arctic stratus cloud deck. The COSMO simulated cloud was also sensitive to changes in the moisture profile. Generally, LWC decreased when RH in the atmospheric layer above the cloud top was decreased. This supports observational and modeling evidence suggesting that the source of water vapor above cloud top is important for the persistence of the liquid layer (Solomon et al., 2011; Sedlar et al., 2012; Morrison et al., 2012). However, in our simulations, introducing a dry layer above the inversion did not cause cloud dissipation. Reducing RH in the sub-cloud layer had only a modest impact on the mixed-phase cloud. Mixed-phase clouds in the Arctic are frequently decoupled from the surface (Sedlar and Shupe, 2014; Sotiropoulou et al., 2014) and therefore do not necessarily rely on a moisture source near the surface to persist.

Increasing the ice crystal concentration to $1\,l^{-1}$ had a moderate influence on the simulated mixed-phase cloud, while an even higher ice crystal concentration of $10\,l^{-1}$ led to glaciation and subsequent dissipation of the cloud. Rogers et al. (2001) found that for thin, low-level stratus clouds, the IN concentration at -15 to -20 °C was around $1\,l^{-1}$. Nevertheless, ice crystal concentrations in the Arctic may vary over 3 orders of magnitude (Morrison et al., 2005) and a maximum IN of $0.25\,l^{-1}$ has been observed in a similar season and geographic region as ASCOS (Bigg, 1996). Considering IN concentration of $0.25\,l^{-1}$ or lower from a past field campaign in the high Arctic (Bigg, 1996) and taking the absence of IN measurements above the instrument detection limit during ASCOS into account, a large increase in ice crystal number concentration to $10\,l^{-1}$, seems an unlikely mechanism responsible for the observed cloud dissipation during ASCOS. Hence, these results suggest that reasonable increases in IN concentrations are not the primary mechanism leading to cloud dissipation for this observed case.
The sensitivity experiments tested here, altering CDNC, ice crystal number concentration, and changing moisture sources to the cloud layer, were designed to mimic changes in the large-scale circulation and advection of air masses with different thermodynamic profiles and aerosol properties. In reality, it is likely that changes in thermodynamical properties and aerosol will happen simultaneously, and that the combination of these processes will control the evolution of the mixed-phase cloud (Kalesse et al., 2016). Nevertheless, we have shown that, independently, dry-air advection above cloud top, ice crystal increase, and CDNC reduction all contribute to a reduction of the liquid condensate layer of a mixed-phased cloud. However, we find that the reduction of CDNC was likely the primary contributor to the dissipation of the observed mixed-phase cloud during this specific case.

To come back to the research questions raised in Chapter 1, the results presented here suggest to draw the following conclusions:

Why does an Arctic mixed-phase cloud dissipate?

- A high ice crystal concentration of $10\,l^{-1}$, which is unrealistic in this environment at the temperatures of the observed cloud, leads to dissipation.
- Dry air advection above the BL inversion has a weakening effect on the cloud liquid layer, but does not lead to dissipation.
- A low CDNC of $2\,cm^{-3}$ leads to dissipation of the observed Arctic mixed-phase cloud.

CHAPTER 7

Microphysical processes in Arctic mixed-phase clouds

The following results are part of a model intercomparison study within the Impact of Biogenic versus Anthropogenic emissions on Clouds and Climate: towards a Holistic UnderStanding (BACCHUS) project[1]. The BACCHUS project is one of three projects in the "Aerosol and Climate" research cluster[2]. It is a European Union funded project (FP7 collaborative project) lead by ETH, Zürich (Switzerland). 21 institutes from the European Union, Switzerland, Norway, and Israel work together to investigate the key processes of aerosol-cloud interactions. The focus is on key regions Amazonian rain forest and Arctic, which are important for Earth's climate. The COSMO model (in the LES setup described in this thesis) participates in a model intercomparison study of an Arctic stratocumulus cloud. A closer look is given at the impact of aerosols and dynamical processes on the stratocumuls cloud (Stevens et al. (2017a, abstract EGU), Stevens et al. (2017b, in prep.)). In this chapter, microphysical processes of Arctic mixed-phase clouds simulated by the COSMO model are analyzed.

[1] http://www.bacchus-env.eu/
[2] http://www.aerosols-climate.org/

7.1 Diagnostic and prognostic CDNC simulations

The simulations described here are similar to the ones in Chapter 6, as the same Arctic mixed-phase cloud system on the same day and time of the Arctic Summer Cloud Ocean Study (ASCOS) campaign is investigated. The setup is similar to the setup in Chapter 6 except for the large scale subsidence (Section 3.4). The simulations are separated into two groups, the diagnostic simulations and the prognostic simulations. The diagnostic simulations have fixed CDNC as in all other simulations in this thesis (Section 3.4). The prognostic simulations are more realistic because for instance collision processes in the cloud are taken into account. All simulations are done with different ice crystal number concentration (ICNC) (0.2, 0.02, 1, and $1\,\mathrm{l}^{-1}$). The simulations with the diagnostic CDNC are comparable to the simulations in Chapter 6. For the simulations with prognostic CDNC, the aerosol size distribution and composition is initialized as a single lognormal mode of ammonium sulfate with a geometric mean diameter of 94 nm and a geometric standard deviation of 1.5. The aerosol number concentration is set to a horizontally and vertically constant of $80\,\mathrm{cm}^{-3}$ or $30\,\mathrm{cm}^{-3}$ at the beginning of the simulation (Table 7.1). The activation of the aerosol depends on the vertical velocity and the supersaturation. In the current prognostic simulations a single updraft velocity which is the vertical velocity at the actual grid point and not a distribution of the the vertical velocity activates the aerosol. The CCN spectrum is computed and finally the activated aerosol number concentration is calculated after Nenes and Seinfeld (2003) and Fountoukis and Nenes (2005). The activated aerosol number concentration depends strongly on the actual vertical velocity (Figure 7.1). At a temperature of -10°C and at a pressure of 900 hPa all particles are activated at a vertical velocity greater than $0.01\,\mathrm{m\,s}^{-1}$. The aerosol concentration of $80\,\mathrm{cm}^{-3}$ needs higher vertical velocities to activate completely than the aerosol concentration of $30\,\mathrm{cm}^{-3}$. To demonstrate differences in diagnostic and prognostic simulations, Figure 7.2 gives an overview of the liquid water content (LWC) and ice water content (IWC) for an ICNC of $0.2\,\mathrm{l}^{-1}$. The upper two mixed-phase cloud simulations show a similar behavior as in Chapter 6. The LWC is decreasing during the CDNC3_ICNC0p2 simulation and the cloud becomes

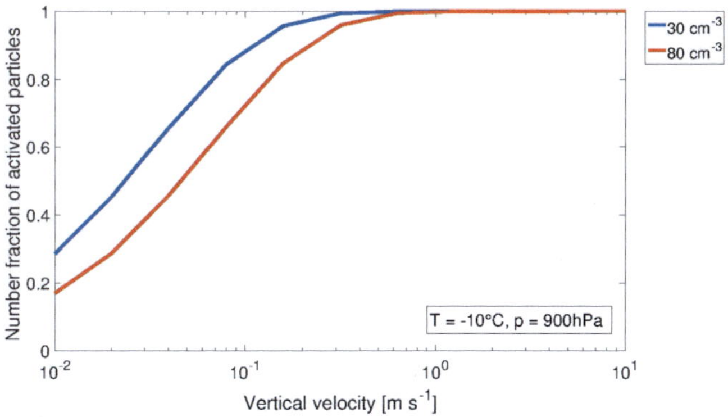

Figure 7.1: Number fraction of activated particles for the horizontally and vertically constant aersol concentration of 80 cm^{-3} and 30 cm^{-3} at -10°C and 900 hPa (courtesy of Markus Karrer, Institute of Meteorology and Climate Research, KIT).

Table 7.1: Overview of the different specifications of the Arctic mixed-phase cloud simulations within the BACCHUS project.

Simulation	Specifications
Diagnostic simulations:	
CDNC30_ICNC1	CDNC = 30 cm^{-3}, ICNC = 1 l^{-1}
CDNC3_ICNC1	CDNC = 3 cm^{-3}, ICNC = 1 l^{-1}
CDNC30_ICNC0p2	CDNC = 30 cm^{-3}, ICNC = 0.2 l^{-1}
CDNC3_ICNC0p2	CDNC = 3 cm^{-3}, ICNC = 0.2 l^{-1}
CDNC30_ICNC0p02	CDNC = 30 cm^{-3}, ICNC = 0.02 l^{-1}
CDNC3_ICNC0p02	CDNC = 3 cm^{-3}, ICNC = 0.02 l^{-1}
CDNC30_noice	CDNC = 30 cm^{-3}, ICNC = 0 l^{-1}
CDNC3_noice	CDNC = 3 cm^{-3}, ICNC = 0 l^{-1}
Prognostic simulations:	
CCN30_ICNC0p2	CCN = 30 cm^{-3}, ICNC = 0.2 l^{-1}
CCN80_ICNC0p2	CCN = 80 cm^{-3}, ICNC = 0.2 l^{-1}
CCN30_ICNC0p02	CCN = 30 cm^{-3}, ICNC = 0.02 l^{-1}
CCN80_ICNC0p02	CCN = 80 cm^{-3}, ICNC = 0.02 l^{-1}
CCN30_noIce	CCN = 30 cm^{-3}, ICNC = 0 l^{-1}
CCN80_noIce	CCN = 80 cm^{-3}, ICNC = 0 l^{-1}

optically thinner, while the mixed-phase cloud layer of the CDNC30_ICNC0p2 simulation persists during the whole simulation time (Figure 7.2a and Figure 7.2b). Compared to these two simulations, the simulations with prognostic CCN show differences (Figure 7.2c and Figure 7.2d). The liquid cloud layer becomes thinner in the CCN=30 cm^{-3} compared to the CCN=80 cm^{-3} simulation, but does

not dissipate. To evaluate how much of the aerosol distribution is activated to droplets, the number density of cloud droplets (QNC) of the four ICNC=$0.2\,l^{-1}$ simulations is studied (Figure 7.3). Comparing Figure 7.3a with Figure 7.3g, both simulations show almost the same amount of QNC around 20 to $25\,cm^{-3}$. $5\,cm^{-3}$ more QNC is seen in the CDNC30_ICNC0p2 simulation towards the end of the simulation. Thus, an aerosol distribution with a maximum value of $30\,cm^{-3}$ produces a QNC of around 20 to $25\,cm^{-3}$, which is in the range of the QNC of the CDNC30_ICNC0p2 simulation. The prognostic CCN=$80\,cm^{-3}$ simulations produces a QNC of around $70\,cm^{-3}$ after two hours of spin-up time. In the prognostic simulation more variation in QNC distribution is seen inside the cloud layer below 1 km (Figure 7.3e, g). The standard deviation of QNC (σ_{QNC}) of the prognostic simulations indicate a higher deviation from the mean at about 700 m than the diagnostic simulations (Figure 7.3b, d, f, h). Thus, higher variations in horizontal direction of QNC are expected in the prognostic simulations than in the diagnostic simulations. Hence, aerosol interactions lead to a more variable QNC.
The cloud liquid water path (LWP) of all 14 different simulations can be divided in three groups (Figure 7.4). The first group includes all three CCN=$80\,cm^{-3}$ simulations. They show the highest values of LWP with maximum values around $65\,g\,m^{-2}$ and appear not to depend on the ICNC amount. Some stronger variations in the LWP time series are seen in the CCN80_ICNC0p2 simulation in the middle of the simulation around 8 h and in the CCN80_ICNC0 simulation towards the end. The minimum LWP is around $40\,g\,m^{-2}$ at 10 h of simulation. The second group is located below group 1 and consists of all three CCN=$30\,cm^{-3}$ and all four CDNC=$30\,cm^{-3}$ simulations. They reach values around $30\,g\,m^{-2}$. No influences of the different ICNCs are seen. More variations in the LWP time series are seen in the prognostic CCN simulations, especially in the CCN30_ICNC0p02 and CCN30_ICNC0 simulations. The third group includes all four CDNC=$3\,cm^{-3}$ simulations. The LWP decreases from around $12\,g\,m^{-2}$ to almost $0\,g\,m^{-2}$ eventually. In brief, the LWP time series depend on the different concentrations of CCN and CDNC.
The ice water path (IWP) of all 14 simulations can be divided into four groups specified by ICNCs (Figure 7.5). The first group includes the two simulations with an ICNC of $1\,l^{-1}$ having the highest IWP values as expected. The CDNC3_ICNC1 simulation shows a fast decrease in IWP after 3 h to almost $0\,g\,m^{-2}$. This Arctic

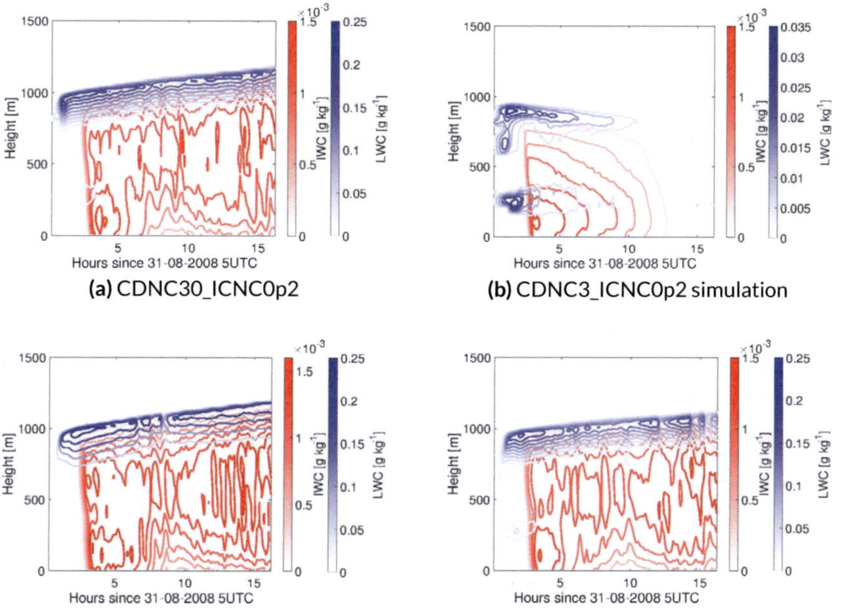

(a) CDNC30_ICNC0p2
(b) CDNC3_ICNC0p2 simulation
(c) CCN80_ICNC0p2 simulation
(d) CCN30_ICNC0p2 simulation

Figure 7.2: Domain average of LWC (blue) and IWC (red) of the CDNCD30_ICNC0p2 simulation **(a)** and the CDNCD3_ICNC0p2 simulation **(b)**, and for the CCN80_ICNC0p2 simulation **(c)** and the CCN30_ICNC0p2 simulation **(d)**.

mixed-phase cloud simulation and the other CDNC=3 cm^{-3} simulations almost completely dissipate towards the end. The CDNC30_ICNC1 simulation shows a minimum amount of IWP after 8 h around 4 $g\,m^{-2}$, while the average IWP is around 5 $g\,m^{-2}$. The second group includes all simulations with an ICNC of 0.2 l^{-1}. The IWP is between 1 $g\,m^{-2}$ and 2 $g\,m^{-2}$ during the whole simulation time and after the 2 h of spin-up time. The IWP of the CDNC3_ICNC0p2 simulation decreases and reaches 0 $g\,m^{-2}$ after 14 h. In the same time range, the LWP of this simulation decreases to values slightly above 0 $g\,m^{-2}$. The third group consists of all simulations with an ICNC of 0.02 l^{-1}. The IWP is above 0 $g\,m^{-2}$ and reaches maximum values around 0.3 $g\,m^{-2}$. The fourth group includes all four simulations with an ICNC of 0 l^{-1}. No ice is formed during these simulations.

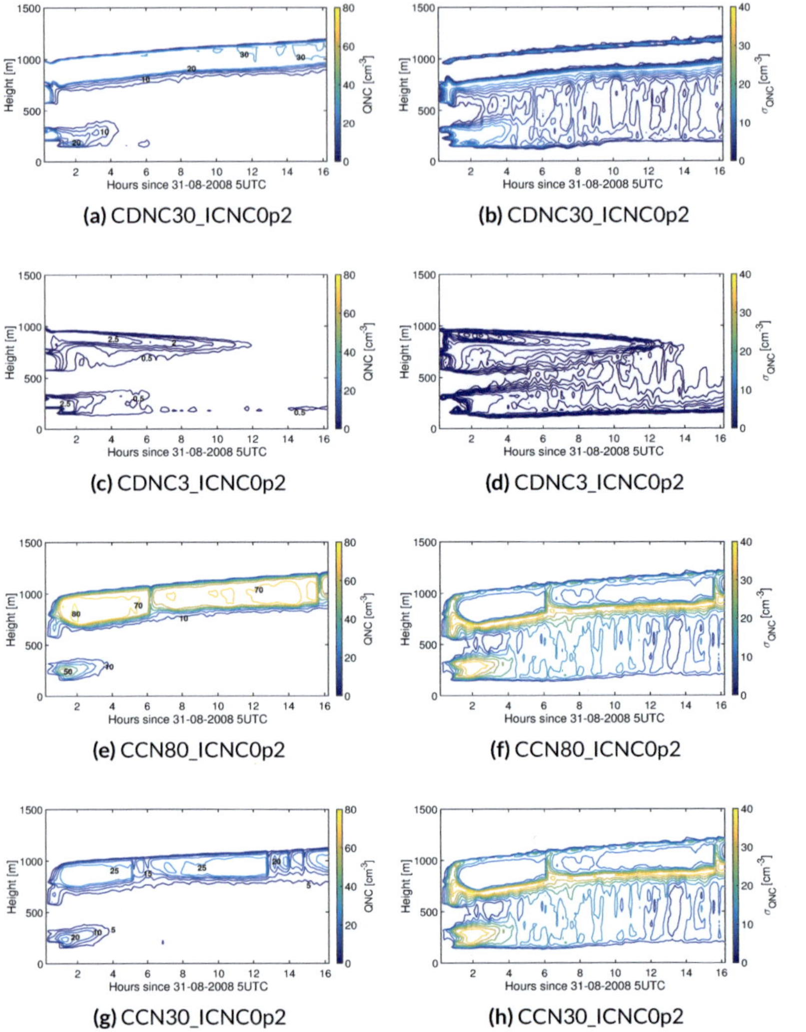

(a) CDNC30_ICNC0p2 (b) CDNC30_ICNC0p2

(c) CDNC3_ICNC0p2 (d) CDNC3_ICNC0p2

(e) CCN80_ICNC0p2 (f) CCN80_ICNC0p2

(g) CCN30_ICNC0p2 (h) CCN30_ICNC0p2

Figure 7.3: Domain averaged QNC and σ_{QNC} of the CDNCD30_ICNC0p2 simulation ((**a**),(**b**)), the CDNCD3_ICNC0p2 ((**c**),(**d**)), the CCN80_ICNC0p2 ((**e**),(**f**)) simulation and the CCN30_ICNC0p2 ((**g**),(**h**)) simulations.

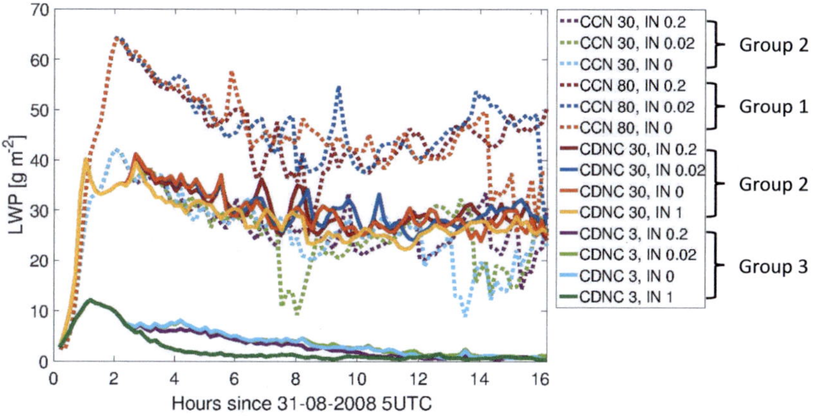

Figure 7.4: The cloud liquid water path (LWP) time series of the 8 diagnostic simulations (solid lines) and the 6 prognostic simulations (dotted lines) for the whole 16 h of simulation time.

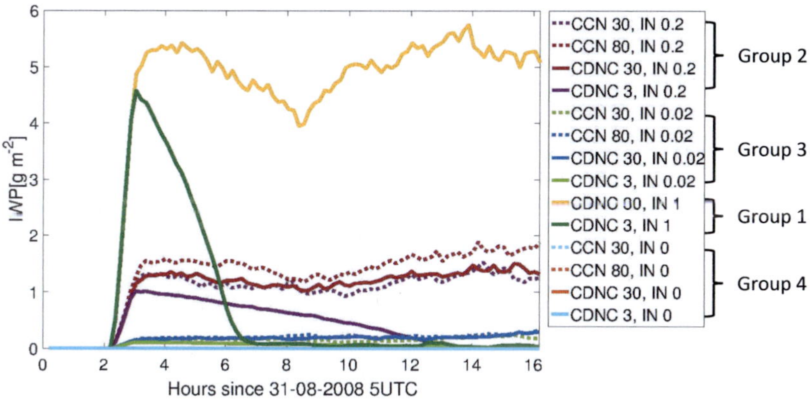

Figure 7.5: The ice water path (IWP) time series of the 8 diagnostic simulations (solid lines) and the 6 prognostic simulations (dotted lines) for the whole 16 h of simulation time.

7.1.1 Microphysical process rates in the simulations with ICNC = 0.2 l^{-1}

The microphysical processes inside Arctic mixed-clouds are analyzed with the following process rates: the autoconversion of cloud droplets to raindrops, the autoconversion of ice particles to snow, the depositional growth of ice and snow, the riming of ice and snow, and the sedimentation rate of rain, ice, and snow (Chapter 2: Figure 2.2 and Figure 2.3). In the model riming occur if certain critical diameters and critical masses are reached. For the ice-cloud- and snow-cloud-riming the diameter of the ice particles have to be larger than 150 μm and the mass of the ice particles have to be larger than 0.01 g kg^{-1}. For the ice-rain- and snow-rain-riming the diameter of the ice particles have to be larger than 100 μm and the mass of the ice particles have to be larger than 0.01 g kg^{-1}. Additionally, the cloud droplets have to be larger than 10 μm and 0.01 g kg^{-1}. The sedimentation processes are sinks of liquid and ice mass in the Arctic mixed-phase cloud. Negative values signify that in a certain vertical level, rain, ice or snow decrease because of the sedimentation process. Positive values signify an increase in rain, ice or snow because of sedimentation from levels above.

The autoconversion of cloud droplets to raindrops is a major process in the diagnostic CDNC simulations in the simulated Arctic mixed-phase cloud (Figure 7.8a). Differences in the autoconversion rate of rain are noticeable in the CDNC3_ICNC0p2 simulation, where the optically thin cloud layer dissipates during the simulation and the autoconversion mainly occurs in the cloud layer around 300 m (Figure 7.8a). The prognostic CCN simulation CCN80_ICNC0p2 shows a lower autoconversion to rain rate in the cloud top layer at about 1 km than in the CDNC30_ICNC0p2 simulation. At the same time, the CCN30_ICNC0p2 simulation shows a stronger autoconversion to rain rate than the CDNC30_ICNC0p2 simulation near the cloud top. The strongest sink of condensate is the sedimentation of rain in the liquid-dominated cloud layer below the cloud top (Figure 7.8e). Furthermore, the CDNC3_ICNC0p2 simulation shows a strong rain-out process in the first 3 h of the simulation because afterwards the mixed-phase cloud dissipates (Figure 7.6).

The autoconversion of ice to snow rate has values in the order of 10^{-15} g cm^{-3} s^{-1} in all four simulations (Figure 7.8i). Compared to the other ice growth and liquid

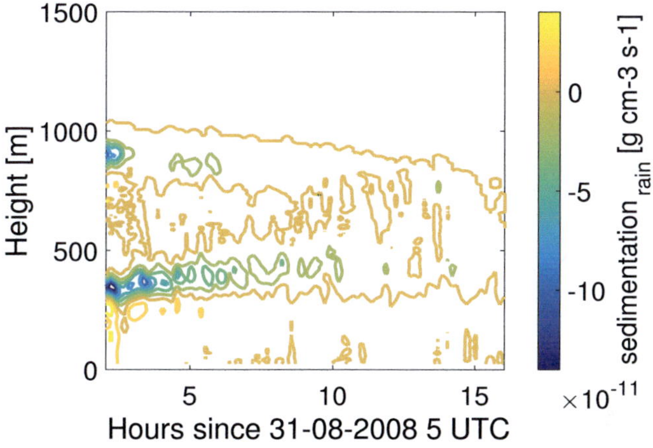

Figure 7.6: Domain average of the sedimentation rate of rain of the diagnostic CDNC3_ICNC0p2 simulation.

growth processes it is not a relevant process. Autoconversion of ice crystals to snow occurs mainly below the liquid-dominated cloud layer towards the surface. The CDNC3_ICNC0p2 simulation has the smallest autoconversion rate of ice crystals to snow because the cloud is dissipating during the simulations.

A further ice growth process is the depositional growth of ice and snow (Figure 7.9d, g). It is an important process in the liquid-dominated cloud layer for all four different simulations. Here, the ice crystals grow on the expense of droplets (Wegener-Bergeron-Findeisen (WBF) process). The depositional growth of ice is negative (e. g. the cloud ice sublimates) near the surface in the prognostic simulations and in the CDNC30_ICNC0p2 simulation. Moreover, the depositional growth of snow is around two orders of magnitude smaller than the depositional growth of ice and occurs mainly in the mixed cloud layer. Near the surface, the depositional growth of snow is negative in all simulations, which indicates that the snow sublimates near the surface (Figure 7.9g).

No riming of ice and snow occurs in the prognostic simulations and in the diagnostic simulations. Thus, according to the results, the depositional growth of ice is an important process for the ice growth.

The negative values of the sedimentation rate of ice and snow show that in certain vertical level, ice or snow decrease because of the sedimentation process (Figure 7.8l and Figure 7.9a). Snow sedimentation is two orders of magnitude smaller than ice sedimentation. The ice sedimentation is positive below the cloud layer, except for the CDNC3_ICNC0p2 simulation. During this simulation all process rates become small after the cloud dissipates.

7.1.2 Microphysical process rates in the simulations with ICNC = $0.02\,\mathrm{l}^{-1}$

The decrease in ICNC to $0.02\,\mathrm{l}^{-1}$ influences the microphysical processes in the Arctic mixed-phase cloud. The autoconversion of cloud droplets to raindrops shows a similar vertical distribution as in the simulations with ICNC to $0.2\,\mathrm{l}^{-1}$ (Figure 7.8b). During the CDNC3_ICNC0p02 simulations, more rain is formed at about 300 m than in the other simulations. The sedimentation of rain occurs most of the time near the cloud top and in the cloud layer in all simulations. The sedimentation rate of the CDNC3_ICNC0p02 simulation is small because the cloud is dissipating during the simulation.
The autoconversion rate from ice to snow is small and around $10^{-18}\,\mathrm{g\,cm^{-3}\,s^{-1}}$ (Figure 7.8j). The process is zero in the CDNC3_ICNC0p02 simulation.
The depositional growth of ice is one order of magnitude smaller than in simulations with higher ICNC (Section 7.1.1, Figure 7.9d, e). At the same time, the depositional growth of snow is around four orders of magnitude smaller and almost zero below the cloud layer and negative near the surface, which means snow is sublimating (Figure 7.9h).
No riming of ice and snow occurs in the prognostic CCN simulations and in the diagnostic CDNC simulations as seen in the ICNC=$0.2\,\mathrm{l}^{-1}$, too. Hence, the depositional growth of ice is the most important process for the ice growth.
In general, the ice growth processes depend on the available amount of ice, which is smaller in these simulations than in the ICNC=$0.2\,\mathrm{l}^{-1}$ simulations. If less ice and snow is present, less ice particles and snow can sediment consequently (Figure 7.8m and Figure 7.9b). The two sedimentation rates, ice and snow sedimentation, are one to four orders of magnitude smaller than the sedimentation rates in the simulations with ICNC=$0.2\,\mathrm{l}^{-1}$ around $10^{-13}\,\mathrm{g\,cm^{-3}\,s^{-1}}$ and $10^{-18}\,\mathrm{g\,cm^{-3}\,s^{-1}}$ (Figure 7.8m and Figure 7.9b).

7.1.3 Microphysical process rates in the simulations with ICNC = 0 l^{-1}

In the simulations with ICNC = 0 l^{-1}, no ice is present in the model and thus no ice or snow can form. The only relevant processes are the liquid phase processes, hence the growth of cloud droplets to rain and the sedimentation of rain.
The autoconversion rates of rain is highest in the CCN30_ICNC0 simulation (Figure 7.8d). The overall vertical distribution of the autoconversion rates is similar to the previous simulations with different ICNC. The autoconversion rate is in the range of the autoconversion rates of the ICNC=0.2 and 0.02 l^{-1}.
Furthermore, the sedimentation of rain serves as a sink to the liquid mass in the cloud and is seen mainly at the cloud top, which is similar to the simulations with ICNC=0.2 and 0.02 l^{-1} (Figure 7.8h).

7.1.4 Microphysical process rates in the simulations with ICNC = 1 l^{-1}

To study the sensitivity of the Arctic mixed-phase cloud to ice, two diagnostic CDNC simulations with an ICNC of 1 l^{-1} are conducted and analyzed.
The autoconversion rate of cloud droplets to rain and the sedimentation of rain are both smaller than in the other diagnostic CDNC simulations (Figure 7.8c, g). The strongest sedimentation of rain is near the cloud top as seen in the previous sections.
The depositional growth of ice is higher than the depositional growth of snow (Figure 7.9c, f). The deposition rate of ice in the CDNC3_ICNC1 simulation is slightly positive at about 300 m in contrast to the CDNC30_ICNC1 simulation, where it is negative. This is due to the second thin cloud layer at about 300 m which is longest seen in the CDNC3 simulations. The autoconversion rate of snow is in the order of 10^{-13} g cm^{-3}s^{-1} and strongest near the surface (Figure 7.8k).
Furthermore, the riming of ice and snow for CDNC30_IN1 is very locally distributed in the cloud layer (Figure 7.7a, b). No snow riming occurs in the CDNC3_IN1 simulation.
The loss of ice mass in the Arctic mixed-phase cloud is shown by the sedimentation rate of ice and snow (Figure 7.8n and Figure 7.9c). The sedimentation of

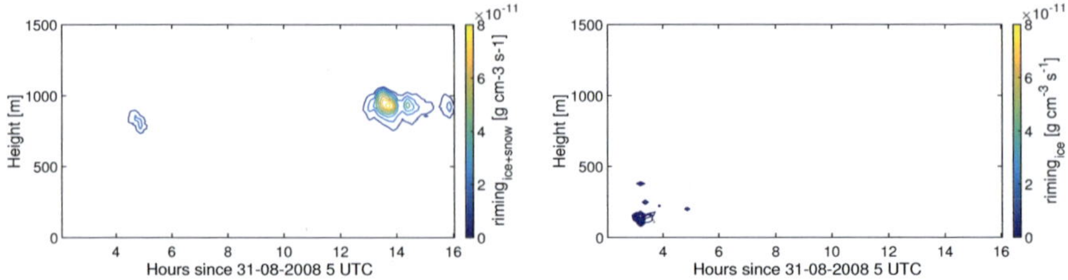

(a) CDNC30_ICNC1 (b) CDNC3_ICNC1

Figure 7.7: Domain averages of the ice and snow riming rate of the diagnostic simulations with CDNC $30\,\mathrm{cm}^{-3}$ (left) and $3\ \mathrm{cm}^{-3}$ (right). All simulations have an ICNC of $1\,\mathrm{l}^{-1}$.

ice is highest near cloud top in the CDNC30_ICNC1 simulation and is around one order of magnitude higher than in the ICNC=$0.2\,\mathrm{l}^{-1}$ simulation (Figure 7.9a). The sedimentation rate of snow is around two orders of magnitudes higher than the other simulations and shows a similar behavior in vertical distribution as in CDNC30_ICNC0p2.

In general, all rates which have to do with ice processes are one or even two orders of magnitude larger compared to the ICNC of $0.2\,\mathrm{l}^{-1}$ simulations, because more ice is initialized in the model and can grow (Figure 7.5). Thus, also the riming process can play a role in the ice growth processes although it is small.

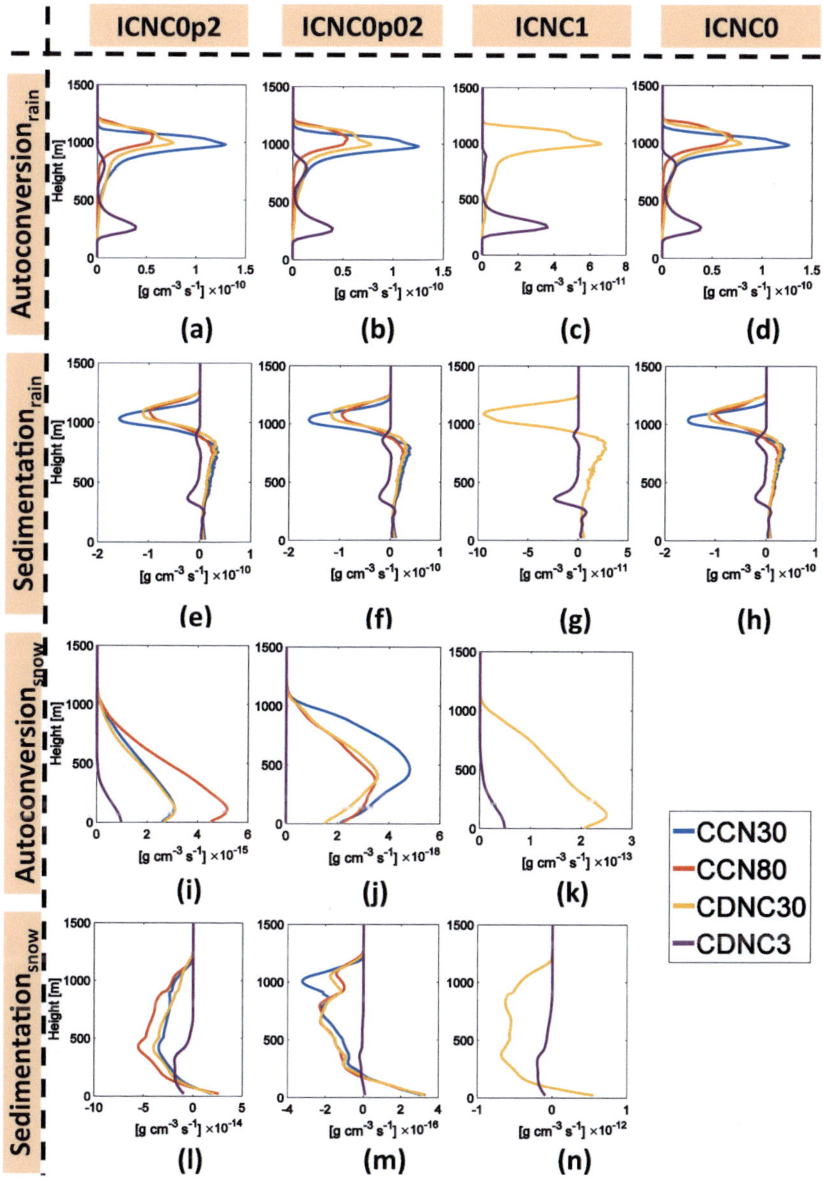

Figure 7.8: Domain and time (2–16 h) averaged autoconversion rate and sedimentation rate of rain and snow for the prognostic and diagnostic simulations with ICNC = 0.2, 0.02, 1, and 0 l^{-1}.

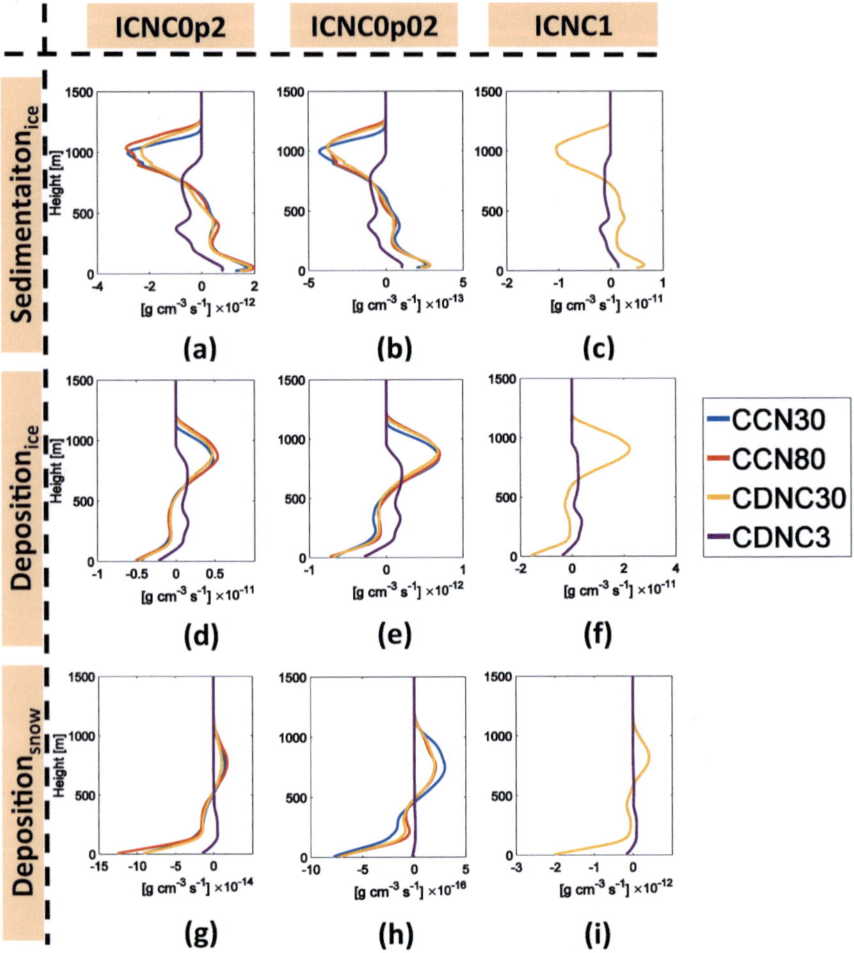

Figure 7.9: Domain and time (2–16 h) averaged sedimenation rate of ice and deposition rate of ice and snow for the prognostic and diagnostic simulations with ICNC = 0.2, 0.02, and 1 l^{-1}.

7.2 Discussion and Conclusions

The sensitivity study of an Arctic mixed-phase cloud during the ASCOS field campaign is part of the BACCHUS project. Insight into the microphysical processes in an Arctic mixed-phase cloud is shown by different simulations. The sensitivity study is divided into two main groups: the diagnostic CDNC simulations and the prognostic CDNC simulations. The sensitivity of the simulated Arctic mixed-phase cloud is analyzed by initially varying the CCN, CDNC, and ICNC (Table 7.1).

The most important process for ice growth in all simulations is the depositional growth of ice. In the simulated Arctic mixed-phase cloud, ice particles grow at the expense of water vapor near the cloud top and in the mixed-phase cloud layer. Further down, near the surface, the ice particles are shrinking by sublimation. The CDNC=3 cm^{-3} simulations behave slightly different. The deposition rate of ice is smaller and distributed more homogeneously in the vertical from at about 900 m to 200 m than in the other simulations. This is probably because the mixed-phase cloud is thinner and two liquid layers are formed. Hence, depositional growth occurs in both liquid layers, the one at about 900 m and the other one at about 300 m. The rate itself is smaller than in all the other simulations because the cloud finally dissipates. The depositional growth of snow and the autoconversion rate of snow are around zero and significantly smaller than the depositional growth of ice and have a small, almost negligible effect on the microphysical cloud processes. The riming process of ice is very locally distributed in the cloud layer. Besides that, riming of ice happens only in the diagnostic CDNC30_ICNC1 and CDNC3_ICNC1 simulations, where it is episodically in the same order of magnitude as the depositional growth of ice. Thus, the WBF process, which describes the ice growth at the expense of cloud droplets is an important process for ice growth in the simulated Arctic mixed-phase cloud.

The main sink of ice particles is the sedimentation of ice in the diagnostic CDNC and prognostic CDNC simulations. It occurs strongest in the cloud layer near cloud top and is also seen in the mixed-phase cloud layer and near the surface. The sedimentation of snow is significantly smaller than the sedimentation of ice and strongest in the near surface layers.

The liquid growth process is described by the autoconversion of cloud droplets to raindrops. In all simulations, the autoconversion rate is highest in the liquid dominated layer near cloud top or in the case of CDNC=3 cm^{-3} in the two liquid layers, where it is strongest in the cloud layer around 300 m. In general, the different ICNCs have only a small effect on the autoconversion rate of rain and a pronounced effect is only seen in the ICNC=1 l^{-1} simulations, where the autoconversion rate is around one order of magnitude smaller than in the other simulations.

Obviously, the diagnostic CDNC=3 cm^{-3} simulation is special because the Arctic mixed-phase cloud has two cloud layers and dissipates towards the end of the simulation. The depositional growth of ice is still an important ice growth process in the cloud layer. Furthermore, the sedimentation of rain is positive around at about 300 m, which shows that a large amount of raindrops from layers above fall into this layer. Finally, the cloud rains out, which leads to smaller microphysical rates than in all the other simulations.

By comparing the sedimentation rate of rain with the sedimentation rate of ice, it is noticeable, that the ice sedimentation is smaller than the rain sedimentation in all simulation except for the highest ICNC simulations (ICNC=1 l^{-1}). Therefore, the smaller concentration of ice leads to the conclusion that the loss of condensate in the cloud is mainly due to the sedimentation of rain. Morrison et al. (2012) described the sedimentation process of ice as an important loss process if the cloud is capped by a dry atmospheric layer. If the cloud is capped by a moisture inversion the moisture can balance the loss by ice sedimentation due to moisture entrainment at cloud top. The current simulations have a moisture inversion above cloud top (Chapter 6, Figure 6.8). Thus, the ice sedimentation can be balanced. Nevertheless, a strong growth of raindrops occurs and leads to be the main sink of condensate of the cloud layer.

To come back to the research questions raised in Chapter 1, the results presented here suggest to draw the following conclusions:

Which microphysical processes are relevant in Arctic mixed-phase clouds?

- Depositional growth of ice is the leading process in the simulated Arctic mixed-phase clouds.
- The autoconversion of cloud droplets to raindrops is strongest near the cloud top of the Artic mixed-phase clouds.
- Varying the initial ICNC does not influence the growth of the liquid droplets in the Arctic mixed-phase clouds.
- Condensate is lost from the cloud layer rather by rain than by ice sedimentation.

CHAPTER 8

Conclusion and Outlook

This thesis showed a detailed analysis of simulated Arctic mixed-phase clouds by dissecting the microphysical properties as well as providing an overview of different influences on Arctic mixed-phase clouds development.The objectives of this thesis were defined in Chapter 1 and will addressed concisely in the following paragraphs. An outlook of possible future extensions is given at the end of this chapter.

Arctic mixed-phase cloud simulations performed by the COSMO model

Field campaigns are important to gather data and to provide a picture of for instance how a cloud evolves or interacts with the surface. However, these measurements are always limited in space and time and never provide a full four-dimensional view of the cloud. For further investigations, a model should reproduce similar mixed-phase clouds. Here, the COSMO model was performing well regarding the main cloud properties such as vertical extent or the liquid water content (LWC). Four Arctic mixed-phase clouds observed during the VERDI campaign were analyzed and compared to observations. Cloud top height and the LWC agreed with the observations. Some differences in the droplet size distribution were found. The simulations showed smaller droplet sizes because of the prescribed shape of the distribution in the two-moment scheme of cloud microphysics. The spatial characteristics of Arctic mixed-phase cloud were analyzed and found to be inhomogeneous both in the observations and in the simulations. To conclude, with the help of measurements, the COSMO model as in the setup used in this thesis, was validated as a basis for the investigations in this thesis.

Influence of different surfaces types and boundary layer (BL) structures on Arctic mixed-phase clouds

The Arctic climate is influenced by the ice covered ocean, which prevents moisture and heat from propagating to the atmosphere. During spring, especially during summer and autumn, the Arctic surface changes and open ocean areas occur. Thus, the question arose if and how these different surface types affects clouds. Furthermore, the Arctic BL structure can be decoupled from or coupled to the surface (Sotiropoulou et al., 2014), which affects the moisture transport from the surface to the atmosphere. To investigate the influences of the surface to Arctic mixed-phase clouds, a sensitivity study was performed with three different surface types (sea ice, open lead, and sea water) and two different BL structures (coupled and decoupled). The simulations showed almost no influence of the surface to the mixed-phase cloud when the BL was decoupled from the surface. If the surface was coupled to the cloud layer during the simulation, the moisture was able to reach the cloud base. This happened in V0429, which is a low mixed-phase cloud with a cloud top at about 590 m. Furthermore, a sensitivity experiment was performed in which the BL was coupled to the surface at the beginning of the simulation by modifying the initial temperature and moisture profile and creating a mixed BL. As a consequence, the moisture reached the cloud base height in all simulations, except for the sea ice coupled simulation of V0515, in which the BL decoupled from the surface during the simulation and kept the moisture near the surface. In general, only small sensible heat flux and latent heat flux occurred during the simulations over the sea ice and open lead surfaces, which consequently had only small influence on the cloud liquid water path (LWP) of the clouds. In contrast, the sea water surface had an influence on the LWP and the ice water path (IWP) in the coupled simulations of V0515 and on the V0429 simulations.

Analysis of the dissipation process of an Arctic mixed-phase cloud

Arctic mixed-phase clouds can persist for several days and have an effect on the surface energy balance. For instance, Morrison et al. (2012) presented a concept why these clouds are so long persistent. Therefore, the question arose how these mixed-phase clouds dissipate. Thus, different mechanisms were analyzed to find the main contributors to dissipation of an Arctic mixed-phase cloud. During the Arctic Summer Cloud Ocean Study (ASCOS) field campaign, a long persistent low-level mixed-phase cloud, which dissipated after some days, was observed and simulations of this cloud were performed with the COSMO model. Three sets of sensitivity experiments were conducted. In the first sensitivity experiment (SensMoist), the moisture profile was modified either below or above the cloud top to mimic dry-air advection. In a second sensitivity experiment (SensIce), the initially set ice crystal number concentration (ICNC) was increased to 1^{-1} and $10\,l^{-1}$. These concentrations were above the expected low values in the Arctic (Bigg, 1996). In the last set of sensitivity experiments (SensCDNC), the fixed cloud droplet number concentration (CDNC) was decreased to $10\,cm^{-3}$ and to $2\,cm^{-3}$. With this set of sensitivity experiments, a change in large-scale advection of air masses was performed to mimic changes in aerosol conditions and thermodynamic profiles. On the whole, the simulations showed that changing the CDNC to $2\,cm^{-3}$ caused the Arctic mixed-phase cloud to dissipate. It is even likely that these different processes occur simultaneously and this model study showed only one contributor at a time. Nevertheless, the main contributor to dissipation was found to be the reduction of CDNC to $2\,cm^{-3}$.

Identifying the main microphysical processes inside Arctic mixed-phase clouds

For a better understanding of the microphysical processes inside Arctic mixed-phase clouds, the same cloud as described in the study about the dissipation processes of an Arctic mixed-phase cloud was investigated. The sensitivity study was divided in two main groups: The first group included diagnostic simulations with a constant CDNC. The initial constant CDNC were $30\,cm^{-3}$ and $3\,cm^{-3}$ and the ICNC were altered to $0\,l^{-1}$, $0.02\,l^{-1}$, $0.2\,l^{-1}$, and $1\,l^{-1}$. The second group included simulations performed with prescribed cloud condensation nuclei

(CCN) and prognostic CDNC. Therefore, the aerosol was initialized as a single lognormal mode of ammonium sulfate with a geometric mean diameter of 94 nm and a geometric standard deviation of 1.5. The aerosol number concentration was set to $30\,\mathrm{cm}^{-3}$ and $80\ \mathrm{cm}^{-3}$ and the ICNC was modified to $0\,\mathrm{l}^{-1}$, $0.02\,\mathrm{l}^{-1}$, and $0.2\,\mathrm{l}^{-1}$. Hence, a set of 14 simulations was performed. The microphysical processes inside Arctic mixed-clouds were analyzed with the following process rates: the autoconversion of cloud droplets to raindrops, the autoconversion of ice particles to snow, the depositional growth of ice and snow, the riming of ice and snow, and the sedimentation rate of rain, ice, and snow. It was found that the warm rain process (autoconversion of cloud droplets to raindrops) was strongest near the cloud top, which is the region where the highest LWC was simulated. Additionally, the strongest sedimentation rate of rain was near the cloud top in all simulations, except for the CDNC=$3\,\mathrm{cm}^{-3}$, where a second layer around 300 m above the surface and below the main cloud layer evolved and finally the cloud dissipated. The sedimentation of rain was also higher than the sedimentation of ice. Thus, the cloud mainly lost condensate via the liquid phase and not the ice phase. Interestingly, the strongest ice phase process is the depositional growth of ice, which indicated that the Wegener-Bergeron-Findeisen (WBF) process was an important process inside the Arctic mixed-phase cloud. Near the surface, the ice particles sublimated. In conclusion, this sensitivity study revealed the microphysical processes inside an Arctic mixed-phase and found that the autoconversion rate of cloud droplets to raindrops as well as the sedimentation of rain played a major role in the cloud layer of the Arctic mixed-phase cloud.

Outlook

Based on the differences in the droplet size distribution discussed in the preceding paragraphs, a further study could investigate the shape of the droplet size distribution and modify it to a more realistic distribution. Useful observations would be measurements of the aerosol number concentrations and size distributions vertically inside an Arctic mixed-phase cloud.

The BL structure is considered to be important in the Arctic. Thus, it would be interesting to see how the BL develops if the Arctic surface temperatures rise. The sea water surface may become more important for the surface energy balance as well as for the moisture transport through the BL. Thus, a mesoscale model study

under climate change conditions could be helpful to assess this question.
The sensitivity study of an Arctic mixed-phase cloud dissected one specific Arctic mixed-phase cloud and elucidated different possible contributors to the dissipation process. Now, it would be interesting to see the main contributors for other clouds for instance at different times of the year. In combination with trajectories, which show the origin of the air masses the relative role of the different contributors to dissipation can be quantified.
The sensitivity study of microphysical processes in Arctic mixed-phase clouds is part of a model intercomparison study within the Impact of Biogenic versus Anthropogenic emissions on Clouds and Climate: towards a Holistic UnderStanding (BACCHUS) project[1]. A comparison of different models is currently being conducted and prepared for publication. Thus, the comparison of different models will help to verify the COSMO model results.

[1] http://www.bacchus-env.eu/

List of Figures

List of Tables

References

Abe, M., T. Nozawa, T. Ogura, and K. Takata, 2016: Effect of retreating sea ice on Arctic cloud cover in simulated recent global warming. *Atmos. Chem. Phys.*, **16 (22)**, 14 343–14 356.

Ackerman, A. S., M. C. vanZanten, B. Stevens, V. Savic-Jovcic, C. S. Bretherton, A. Chlond, J.-C. Golaz, H. Jiang, M. Khairoutdinov, S. K. Krueger, D. C. Lewellen, A. Lock, C.-H. Moeng, K. Nakamura, M. D. Petters, J. R. Snider, S. Weinbrecht, and M. Zulauf, 2009: Large-eddy simulations of a drizzling, stratocumulus-topped marine boundary layer. *Mon. Wea. Rev.*, **137 (3)**, 1083–1110.

Bär, J., 2012: Charakterisierung der synoptischen Situation während der Messkampagne VERDI (April/Mai 2012). B.s. thesis, University of Leipzig.

Bergeron, T., 1935: On the physics of cloud and precipitation. *Proc. 5th Assembly U.G.G.I. Lisbon*, **2**, 156 – 178.

Bigg, E. K., 1996: Ice forming nuclei in the high Arctic. *Tellus B*, **48 (2)**, 223–233.

Birch, C. E., I. M. Brooks, M. Tjernström, M. D. Shupe, T. Mauritsen, J. Sedlar, A. P. Lock, P. Earnshaw, P. O. G. Persson, S. F. Milton, and C. Leck, 2012: Modelling atmospheric structure, cloud and their response to CCN in the central Arctic: ASCOS case studies. *Atmos. Chem. Phys.*, **12 (7)**, 3419–3435.

Brown, A. R., S. H. Derbyshire, and P. J. Mason, 1994: Large-eddy simulation of stable atmospheric boundary layers with a revised stochastic subgrid model. *Quarterly Journal of the Royal Meteorological Society*, **120 (520)**, 1485–1512.

Brümmer, B., 1999: Roll and cell convection in wintertime Arctic cold-air outbreaks. *J. Atmos. Sci.*, **56 (15)**, 2613–2636.

Burnet, F. and J.-L. Brenguier, 2007: Observational study of the entrainment-mixing process in warm convective clouds. *J. Atmos. Sci.*, **64 (6)**, 1995–2011.

Curry, J. A., J. L. Schramm, W. B. Rossow, and D. Randall, 1996: Overview of Arctic cloud and radiation characteristics. *J. Climate*, **9 (8)**, 1731–1764.

Davis, A., R. Cahalan, D. Spinehirne, M. McGill, and S. Love, 1999: Off-beam lidar: an emerging technique in cloud remote sensing based on radiative green-function theory in the diffusion domain. *Phys. Chem. Earth Pt. B*, **24 (3)**, 177 – 185.

Dee, D. P., S. M. Uppala, A. J. Simmons, P. Berrisford, P. Poli, S. Kobayashi, U. Andrae, M. A. Balmaseda, G. Balsamo, P. Bauer, P. Bechtold, A. C. M. Beljaars, L. van de Berg, J. Bidlot, N. Bormann, C. Delsol, R. Dragani, M. Fuentes, A. J. Geer, L. Haimberger, S. B. Healy, H. Hersbach, E. V. Hólm, L. Isaksen, P. Kållberg, M. Köhler, M. Matricardi, A. P. McNally, B. M. Monge-Sanz, J.-J. Morcrette, B.-K. Park, C. Peubey, P. de Rosnay, C. Tavolato, J.-N. Thépaut, and F. Vitart, 2011: The ERA-Interim reanalysis: configuration and performance of the data assimilation system. *Q.J.R. Meteorol. Soc.*, **137 (656)**, 553–597.

Durran, D. R. and J. B. Klemp, 1982: On the effects of moisture on the Brunt-Väisälä frequency. *J. Atmos. Sci.*, **39 (10)**, 2152–2158.

Findeisen, W., 1938: Die kolloidmeteorologischen Vorgänge bei der Niederschlagsbildung. *Met. Z.*, **(55)**, 121 – 133.

Fountoukis, C. and A. Nenes, 2005: Continued development of a cloud droplet formation parameterization for global climate models. *J. Geophys. Res. Atmos.*, **110 (D11)**.

Garrett, T. J., C. Zhao, and P. C. Novelli, 2010: Assessing the relative contributions of transport efficiency and scavenging to seasonal variability in Arctic aerosol. *Tellus B*, **62 (3)**, 190–196.

Gerber, H., G. Frick, S. P. Malinowski, H. Jonsson, D. Khelif, and S. K. Krueger, 2013: Entrainment rates and microphysics in POST stratocumulus. *J. Geophys. Res. Atmos.*, **118 (21)**, 12,094–12,109.

Heintzenberg, J. and C. Leck, 2012: The summer aerosol in the central Arctic 1991–2008: did it change or not? *Atmos. Chem. Phys.*, **12 (9)**, 3969–3983.

Heintzenberg, J., C. Leck, W. Birmli, B. Wehner, B., M. Tjernström, and A. Wiedensohler, 2006: Aerosol number-size distributions during clear and fog periods in the summer high Arctic: 1991, 1996 and 2001. *Tellus B*, **58 (1)**, 41–50.

Herzog, H.-J., U. Schubert, G. Vogel, A. Fidler, and R. Kirchner, 2002a: LLM - the high-resolving nonhydrostatic simulation model in the DWD - project LITFASS.

Part I modelling technique and simulation method. COSMO Technical Report No. 4, Deutscher Wetterdienst, D-63004 Offenbach, Germany.

Herzog, H.-J., G. Vogel, and U. Schubert, 2002b: LLM – a nonhydrostatic model applied to high-resolving simulations of turbulent fluxes over heterogeneous terrain. *Theor. Appl. Climatol.*, **73 (1)**, 67–86.

Hobbs, P. V. and A. L. Rangno, 1998: Microstructures of low and middle-level clouds over the beaufort sea. *Q.J.R. Meteorol. Soc.*, **124 (550)**, 2035–2071.

Intrieri, J. M., C. W. Fairall, M. D. Shupe, P. O. G. Persson, E. L. Andreas, P. S. Guest, and R. E. Moritz, 2002: An annual cycle of Arctic surface cloud forcing at SHEBA. *J. Geophys. Res.*, **107 (C10)**, 8039–.

Kalesse, H., G. de Boer, A. Solomon, M. Oue, M. Ahlgrimm, D. Zhang, M. D. Shupe, E. Luke, and A. Protat, 2016: Understanding rapid changes in phase partitioning between cloud liquid and ice in stratiform mixed-phase clouds: An Arctic case study. *Mon. Wea. Rev.*, **144 (12)**, 4805–4826.

Kay, J. and A. Gettelman, 2009: Cloud influence on and response to seasonal Arctic sea ice loss. *J. Geophys. Res.*, **114 (18)**, –.

Khain, A. P., K. D. Beheng, A. Heymsfield, A. Korolev, S. O. Krichak, Z. Levin, M. Pinsky, V. Phillips, T. Prabhakaran, A. Teller, S. C. van den Heever, and J.-I. Yano, 2015: Representation of microphysical processes in cloud-resolving models: spectral (bin) microphysics vs. bulk parameterization. *Rev. Geophys.*, 2014RG000 468–.

Klingebiel, M., A. de Lozar, S. Molleker, R. Weigel, A. Roth, L. Schmidt, J. Meyer, A. Ehrlich, R. Neuber, M. Wendisch, and S. Borrmann, 2015: Arctic low-level boundary layer clouds: in situ measurements and simulations of mono- and bimodal supercooled droplet size distributions at the top layer of liquid phase clouds. *Atmos. Chem. Phys.*, **15 (2)**, 617–631.

Köhler, H., 1936: The nucleus in and the growth of hygroscopic droplets. *Transacttions of the Faraday Society*, **32**, 1152–1161.

Korolev, A., 2007: Limitations of the Wegener-Bergeron-Findeisen mechanism in the evolution of mixed-phase clouds. *J. Atmos. Sci.*, **64 (9)**, 3372–3375.

Korolev, A. V. and I. P. Mazin, 2003: Supersaturation of water vapor in clouds. *J. Atmos. Sci.*, **60 (24)**, 2957–2974.

Lamb, D. and J. Verlinde, 2011: *Physics and chemistry of clouds*. Cambridge University Press, New York.

Langhans, W., J. Schmidli, and B. Szintai, 2012: A smagorinsky-lilly turbulence closure for COSMO-LES: Implementation and comparison to ARPS. *COSMO newsletter*, **No. 12**, 20–31.

Leck, C. and E. Svensson, 2015: Importance of aerosol composition and mixing state for cloud droplet activation over the Arctic pack ice in summer. *Atmos.Chem. Phys.*, **15 (5)**, 2545–2568.

Lilly, D. K., 1962: On the numerical simulation of buoyant convection. *Tellus*, **14 (2)**, 148–172.

Loewe, K., A. M. L. Ekman, M. Paukert, J. Sedlar, M. Tjernström, and C. Hoose, 2016: Modelling micro- and macrophysical contributors to the dissipation of an Arctic mixed-phase cloud during the Arctic summer cloud ocean study (ASCOS). *Atmos. Chem. Phys. Discuss.*, **2016**, 1–22.

Manabe, S. and R. J. Stouffer, 1980: Sensitivity of a global climate model to an increase of CO2 concentration in the atmosphere. *J. Geophys. Res.*, **85 (C10)**, 5529–5554.

Martin, M., R. Y.-W. Chang, B. Sierau, S. Sjogren, E. Swietlicki, J. P. D. Abbatt, C. Leck, and U. Lohmann, 2011: Cloud condensation nuclei closure study on summer Arctic aerosol. *Atmos. Chem. Phys.*, **11 (22)**, 11 335–11 350.

Mauritsen, T., J. Sedlar, M. Tjernström, C. Leck, M. Martin, M. Shupe, S. Sjogren, B. Sierau, P. O. G. Persson, I. M. Brooks, and E. Swietlicki, 2011: An arctic CCN-limited cloud-aerosol regime. *Atmos.Chem. Phys.*, **11 (1)**, 165–173.

McFarquhar, G. M., S. Ghan, J. Verlinde, A. Korolev, J. W. Strapp, B. Schmid, J. M. Tomlinson, M. Wolde, S. D. Brooks, D. Cziczo, M. K. Dubey, J. Fan, C. Flynn, I. Gultepe, J. Hubbe, M. K. Gilles, A. Laskin, P. Lawson, W. R. Leaitch, P. Liu, X. Liu, D. Lubin, C. Mazzoleni, A.-M. Macdonald, R. C. Moffet, H. Morrison, M. Ovchinnikov, M. D. Shupe, D. D. Turner, S. Xie, A. Zelenyuk, K. Bae, M. Freer, and A. Glen, 2011: Indirect and semi-direct aerosol campaign. *Bull*, **92 (2)**, 183–201.

Mironov, D., 2008: Parameterization of lakes in numerical weather prediction. Description of a lake model. COSMO Technical Report No. 11, Deutscher Wetterdienst, D-63004 Offenbach, Germany.

Mironov, D. and B. Ritter, 2004: Testing the new ice model for the global NWP system GME of the German weather service. Technical documentation 1220, WMO.

Morrison, H., G. de Boer, G. Feingold, J. Harrington, M. D. Shupe, and K. Sulia, 2012: Resilience of persistent Arctic mixed-phase clouds. *Nat. Geosci.*, **5 (1)**, 11–17.

Morrison, H., M. D. Shupe, J. O. Pinto, and J. A. Curry, 2005: Possible roles of ice nucleation mode and ice nuclei depletion in the extended lifetime of Arctic mixed-phase clouds. *Geophys. Res. Lett.*, **32 (18)**, L18 801–.

Morrison, H., P. Zuidema, A. S. Ackerman, A. Avramov, G. de Boer, J. Fan, A. M. Fridlind, T. Hashino, J. Y. Harrington, Y. Luo, M. Ovchinnikov, and B. Shipway, 2011: Intercomparison of cloud model simulations of Arctic mixed-phase boundary layer clouds observed during SHEBA/FIRE-ACE. *J. Adv. Model. Earth Syst.*, **3 (2)**, M05 001–.

Nenes, A. and J. H. Seinfeld, 2003: Parameterization of cloud droplet formation in global climate models. *J. Geophys. Res. Atmos.*, **108 (D14)**.

Nygård, T., T. Valkonen, and T. Vihma, 2014: Characteristics of Arctic low-tropospheric humidity inversions based on radio soundings. *Atmos. Chem. Phys.*, **14 (4)**, 1959–1971.

Ovchinnikov, M., A. S. Ackerman, A. Avramov, A. Cheng, J. Fan, A. M. Fridlind, S. Ghan, J. Harrington, C. Hoose, A. Korolev, G. M. McFarquhar, H. Morrison, M. Paukert, J. Savre, B. J. Shipway, M. D. Shupe, A. Solomon, and K. Sulia, 2014: Intercomparison of large-eddy simulations of Arctic mixed-phase clouds: Importance of ice size distribution assumptions. *J. Adv. Model. Earth Syst.*, **6 (1)**, 223–248.

Paukert, M. and C. Hoose, 2014: Modeling immersion freezing with aerosol-dependent prognostic ice nuclei in Arctic mixed-phase clouds. *J. Geophys. Res. Atmos.*, **119 (14)**, 9073–9092.

Penner, J. E., 2004: Climate change: The cloud conundrum. *Nature*, **432 (7020)**, 962–963.

Pinto, J. O., 1998: Autumnal mixed-phase cloudy boundary layers in the Arctic. *J. Atmos. Sci.*, **55 (11)**, 2016–2038.

Pruppacher, H. R. and J. D. Klett, 1997: *Microphysics of cloud and precipitation*, Vol. 18. 2d ed., Kluwer Academic Publishers.

Rangno, A. L. and P. V. Hobbs, 2001: Ice particles in stratiform clouds in the Arctic and possible mechanisms for the production of high ice concentrations. *J. Geophys. Res.*, **106 (D14)**, 15 065–15 075.

Ritter, B. and J.-F. Geleyn, 1992: A comprehensive radiation scheme for numerical weather prediction models with potential applications in climate simulations. *Mon. Wea. Rev.*, **120 (2)**, 303–325.

Rogers, D. C., P. J. DeMott, and S. M. Kreidenweis, 2001: Airborne measurements of tropospheric ice-nucleating aerosol particles in the Arctic spring. *J. Geophys. Res.*, **106**, 15.

Sato, K., J. Inoue, Y. Kodama, and J. E. Overland, 2012: Impact of Arctic sea-ice retreat on the recent change in cloud-base height during autumn. *Geophys.*, **39 (10)**, –.

Savre, J., A. Ekman, G. Svensson, and M. Tjernström, 2014: Large-eddy simulations of an Arctic mixed-phase stratiform cloud observed during ISDAC : Sensitivity to moisture aloft, surface fluxes and large-scale forcing. *Q.J.R. Meteorol. Soc.*

Schäfer, M., E. Bierwirth, A. Ehrlich, F. Heyner, and M. Wendisch, 2013: Retrieval of cirrus optical thickness and assessment of ice crystal shape from ground-based imaging spectrometry. *Atmos. Meas. Tech.*, **6 (8)**, 1855–1868.

Schäfer, M., E. Bierwirth, A. Ehrlich, E. Jäkel, and M. Wendisch, 2015: Airborne observations and simulations of three-dimensional radiative interactions between Arctic boundary layer clouds and ice floes. *Atmos. Chem. Phys.*, **15 (14)**, 8147–8163.

Schäfer, M., E. Bierwirth, A. Ehrlich, E. Jäkel, F. Werner, and M. Wendisch, 2017: Directional, horizontal inhomogeneities of cloud optical thickness fields retrieved from ground-based and airbornespectral imaging. *Atmos. Chem. Phys.*, **17 (3)**, 2359–2372.

Schättler, U., G. Doms, and C. Schraff, 2015: A description of the non-hydrostatic regional cosmo-model, part VII: user's guide.

Schweiger, A. J., R. W. Lindsay, S. Vavrus, and J. A. Francis, 2008: Relationships between Arctic sea ice and clouds during autumn. *J. Climate*, **21 (18)**, 4799–4810.

Sedlar, J. and M. D. Shupe, 2014: Characteristic nature of vertical motions observed in Arctic mixed-phase stratocumulus. *Atmos.Chem. Phys.*, **14 (7)**, 3461–3478.

Sedlar, J., M. D. Shupe, and M. Tjernström, 2012: On the relationship between thermodynamic structure and cloud top, and its climate significance in the Arctic. *J. Climate*, **25 (7)**, 2374–2393.

Sedlar, J., M. Tjernström, T. Mauritsen, M. Shupe, I. Brooks, P. Persson, C. Birch, C. Leck, A. Sirevaag, and M. Nicolaus, 2011: A transitioning Arctic surface energy budget: the impacts of solar zenith angle, surface albedo and cloud radiative forcing. *Clim. Dyn.*, **37 (7-8)**, 1643–1660–.

Seifert, A. and K. D. Beheng, 2001: A double-moment parameterization for simulating autoconversion, accretion and selfcollection. *Atmos. Res.*, **59-60**, 265 – 281.

———, 2006: A two-moment cloud microphysics parameterization for mixed-phase clouds. Part 1: Model description. *Meteorol. Atmos. Phys.*, **92 (1-2)**, 45–66.

Seinfeld, J. H. and S. N. Pandis, 2016: *Atmospheric chemistry and physics From Air pollution to climate change*. 3d ed., John Wiley & Sons, Inc.

Serreze, M. C., A. P. Barrett, J. C. Stroeve, D. N. Kindig, and M. M. Holland, 2009: The emergence of surface-based Arctic amplification. *The Cryosphere*, **3 (1)**, 11–19.

Serreze, M. C. and R. G. Barry, 2011: Processes and impacts of Arctic amplification: A research synthesis. *Global and Planetary Change*, **77 (1-2)**, 85 – 96.

Serreze, M. C. and J. Stroeve, 2015: Arctic sea ice trends, variability and implications for seasonal ice forecasting. *Philosophical Transactions of the Royal Society of London A: Mathematical, Physical and Engineering Sciences*, **373 (2045)**.

Shupe, M. D., 2007: A ground-based multisensor cloud phase classifier. *Geophys. Res. Lett.*, **34 (22)**.

Shupe, M. D. and J. M. Intrieri, 2004: Cloud radiative forcing of the Arctic surface: The influence of cloud properties, surface albedo, and solar zenith angle. *J. Climate*, **17 (3)**, 616–628.

Shupe, M. D., P. Kollias, P. O. G. Persson, and G. M. McFarquhar, 2008: Vertical motions in Arctic mixed-phase stratiform clouds. *J. Atmos. Sci.*, **65 (4)**, 1304–1322.

Shupe, M. D., S. Y. Matrosov, and T. Uttal, 2006: Arctic mixed-phase cloud properties derived from surface-based sensors at SHEBA. *J. Atmos. Sci.*, **63 (2)**, 697–711.

Shupe, M. D., P. O. G. Persson, I. M. Brooks, M. Tjernström, J. Sedlar, T. Mauritsen, S. Sjogren, and C. Leck, 2013: Cloud and boundary layer interactions over the Arctic sea ice in late summer. *Atmos. Chem. Phys.*, **13 (18)**, 9379–9399.

Shupe, M. D., T. Uttal, and S. Y. Matrosov, 2005: Arctic cloud microphysics retrievals from surface-based remote sensors at SHEBA. *J. Appl. Meteor.*, **44 (10)**, 1544–1562.

Shupe, M. D., V. P. Walden, E. Eloranta, T. Uttal, J. R. Campbell, S. M. Starkweather, and M. Shiobara, 2011: Clouds at Arctic atmospheric observatories. Part I: Occurrence and macrophysical properties. *J. Appl. Meteor. Climatol.*, **50 (3)**, 626–644.

Smagorinsky, J., 1963: General circulation experiments with the primitive equations. *Mon. Wea. Rev.*, **91 (3)**, 99–164.

Solomon, A., M. D. Shupe, O. Persson, H. Morrison, T. Yamaguchi, P. M. Caldwell, and G. de Boer, 2013: The sensitivity of springtime Arctic mixed-phase stratocumulus clouds to surface-layer and cloud-top inversion-layer moisture sources. *J. Atmos. Sci.*, **71 (2)**, 574–595.

Solomon, A., M. D. Shupe, P. O. G. Persson, and H. Morrison, 2011: Moisture and dynamical interactions maintaining decoupled Arctic mixed-phase stratocumulus in the presence of a humidity inversion. *Atmos. Chem. Phys.*, **11 (19)**, 10 127–10 148.

Sotiropoulou, G., J. Sedlar, R. Forbes, and M. Tjernström, 2016: Summer Arctic clouds in the ECMWF forecast model: an evaluation of cloud parametrization schemes. *Q.J.R. Meteorol. Soc.*, **142 (694)**, 387–400.

Sotiropoulou, G., J. Sedlar, M. Tjernström, M. D. Shupe, I. M. Brooks, and P. O. G. Persson, 2014: The thermodynamic structure of summer Arctic stratocumulus and the dynamic coupling to the surface. *Atmos. Chem. Phys.*, **14 (22)**, 12 573–12 592.

Stevens, B., C.-H. Moeng, A. S. Ackerman, C. S. Bretherton, A. Chlond, S. de Roode, J. Edwards, J.-C. Golaz, H. Jiang, M. Khairoutdinov, M. P. Kirkpatrick, D. C. Lewellen, A. Lock, F. Müller, D. E. Stevens, E. Whelan, and P. Zhu, 2005: Evaluation of large-eddy simulations via observations of nocturnal marine stratocumulus. *Mon. Wea. Rev.*, **133 (6)**, 1443–1462.

Stevens, R. G., C. Dearden, A. Possner, K. Loewe, T. Raatikainen, A. Dimitrelos, G. Eirund, A. A. Hill, B. J. Shipway, P. Connolly, A. Ekman, C. Hoose, K. S.

Carslaw, and P. Field, 2017a: Model intercomparison of CCN-limited arctic clouds during ASCOS. Abstract.

———, 2017b: A model intercomparison of CCN-limited clouds in the Arctic. In prep.

Stroeve, J., M. Serreze, S. Drobot, S. Gearheard, M. Holland, J. Maslanik, W. Meier, and T. Scambos, 2008: Arctic sea ice extent plummets in 2007. *EOS, Trans. Am. Geophys. Union*, **89 (2)**, 13–14.

Szczap, F., H. Isaka, M. Saute, B. Guillemet, and A. Ioltukhovski, 2000: Effective radiative properties of bounded cascade nonabsorbing clouds: Definition of the equivalent homogeneous cloud approximation. *J. Geophys. Res. Atmos.*, **105 (D16)**, 20 617–20 633.

Tjernström, M., C. E. Birch, I. M. Brooks, M. D. Shupe, P. O. G. Persson, J. Sedlar, T. Mauritsen, C. Leck, J. Paatero, M. Szczodrak, and C. R. Wheeler, 2012: Meteorological conditions in the central Arctic summer during the arctic summer cloud ocean study (ASCOS). *Atmos. Chem. Phys.*, **12 (15)**, 6863–6889.

Tjernström, M. and R. G. Graversen, 2009: The vertical structure of the lower Arctic troposphere analysed from observations and the ERA-40 reanalysis. *Quarterly Journal of the Royal Meteorological Society*, **135 (639)**, 431–443.

Tjernström, M., C. Leck, C. Birch, J. Bottenheim, B. Brooks, I. Brooks, L. Bäcklin, R.-W. Chang, G. De Leeuw, L. Di Liberto, S. De La Rosa, E. Granath, M. Graus, A. Hansel, J. Heintzenberg, A. Held, A. Hind, P. Johnston, J. Knulst, M. Martin, P. Matrai, T. Mauritsen, M. Müller, S. Norris, M. Orellana, D. Orsini, J. Paatero, P. Persson, Q. Gao, C. Rauschenberg, Z. Ristovski, J. Sedlar, M. Shupe, B. Sierau, A. Sirevaag, S. Sjogren, O. Stetzer, E. Swietlicki, M. Szczodrak, P. Vaattovaara, N. Wahlberg, M. Westberg, and C. Wheeler, 2014: The Arctic summer cloud ocean study (ASCOS): Overview and experimental design. *Atmos. Chem. Phys.*, **14 (6)**, 2823–2869.

Uttal, T., J. A. Curry, M. G. Mcphee, D. K. Perovich, R. E. Moritz, J. A. Maslanik, P. S. Guest, H. L. Stern, J. A. Moore, R. Turenne, A. Heiberg, M. C. Serreze, D. P. Wylie, O. G. Persson, C. A. Paulson, C. Halle, J. H. Morison, P. A. Wheeler, A. Makshtas, H. Welch, M. D. Shupe, J. M. Intrieri, K. Stamnes, R. W. Lindsey, R. Pinkel, W. S. Pegau, T. P. Stanton, and T. C. Grenfeld, 2002: Surface heat budget of the Arctic ocean. *Bull. Amer. Meteor. Soc.*, **83 (2)**, 255–275.

Vaughan, D., J. Comiso, I. Allison, J. Carrasco, G. Kaser, R. Kwok, P. Mote, T. Murray, F. Paul, J. Ren, E. Rignot, O. Solomina, K. Steffen, and T. Zhang,

2013: *Observations: Cryosphere. In: Climate Change 2013: The Physical Science Basis. Contribution of Working Group I to the Fifth Assessment Report of the Intergovernmental Panel on Climate Change*. Cambridge University Press, Cambridge, United Kingdom and New York, NY, USA.

Verlinde, J., J. Y. Harrington, V. T. Yannuzzi, A. Avramov, S. Greenberg, S. J. Richardson, C. P. Bahrmann, G. M. McFarquhar, G. Zhang, N. Johnson, M. R. Poellot, J. H. Mather, D. D. Turner, E. W. Eloranta, D. C. Tobin, R. Holz, B. D. Zak, M. D. Ivey, A. J. Prenni, P. J. DeMott, J. S. Daniel, G. L. Kok, K. Sassen, D. Spangenberg, P. Minnis, T. P. Tooman, M. Shupe, A. J. Heymsfield, and R. Schofield, 2007: The mixed-phase Arctic cloud experiment. *Bull. Amer. Meteor. Soc.*, **88 (2)**, 205–221.

Vihma, T., J. Screen, M. Tjernström, B. Newton, X. Zhang, V. Popova, C. Deser, M. Holland, and T. Prowse, 2016: The atmospheric role in the Arctic water cycle: A review on processes, past and future changes, and their impacts. *J. Geophys. Res. Biogeosci.*, 2015JG003132.

Vochezer, P., E. Järvinen, R. Wagner, P. Kupiszewski, T. Leisner, and M. Schnaiter, 2016: In situ characterization of mixed phase clouds using the Small Ice Detector and the Particle Phase Discriminator. *Atmos. Meas. Tech.*, **9 (1)**, 159–177.

Wallace, J. M. and P. V. Hobbs, 2006: *Atmospheric science: an introductory survey*. Academic Press, Amsterdam.

Wegener, A., 1911: *Thermodynamik der Atmosphäre*. Barth, Leipzig.

Wesslén, C., M. Tjernström, D. H. Bromwich, G. de Boer, A. M. L. Ekman, L.-S. Bai, and S.-H. Wang, 2014: The Arctic summer atmosphere: an evaluation of reanalyses using ASCOS data. *Atmos. Chem. Phys.*, **14 (5)**, 2605–2624.

Westwater, E. R., Y. Han, M. D. Shupe, and S. Y. Matrosov, 2001: Analysis of integrated cloud liquid and precipitable water vapor retrievals from microwave radiometers during the surface heat budget of the arctic ocean project. *J. Geophys. Res. Atmos.*, **106 (D23)**, 32 019–32 030.

Willmes, S. and G. Heinemann, 2016: Sea-ice wintertime lead frequencies and regional characteristics in the Arctic, 2003-2015. *Remote Sens.*, **8(1)**, 4.

Acknowledgments

Many people accompanied my PhD student time and contributed to this thesis in different ways during the last years.
First, I would like to thank my supervisor Prof. Corinna Hoose. She introduced me to the modeling world and gave me the opportunity to work in this very interesting field. I appreciate our regular meetings and discussions in a very friendly and motivating atmosphere. Thank you for your encouragement and for making it possible to join several conferences, workshops, and soft skill courses within KHYS and GRACE.
I am very thankful that Prof. Leisner accepted to referee the thesis, takes the time to review it, and introduced me to the art of lobster eating.
I thank the Graduate School for Climate and Environment (GRACE) for funding my research stay at Stockholm University. In particular, I would like to thank Prof. Annica Ekman for supervising me during my time in Stockholm. She always took the time to discuss simulations with me and contributed good ideas on how to proceed. She also introcuded me to the BACCHUS community putting the COSMO simulations in a wider context and contributing to model validation. I thank the BACCHUS community for letting me participate in this great project. I would also like to thank Prof. Michael Tjernström and Dr. Joseph Sedlar for sharing their experience in Arctic science and ASCOS data with me. I thank Lena Frey, Aitor Aldama Campino, Maartje Kuilman, Waheed Iqbal, and Xiang-Yu Li for the warm welcome in Stockholm. I thank the VERDI community for providing me observational data, especially Dr. Paul Vochezer, Dr. Martin Schnaiter, and Dr. Michael Schäfer. I am very greatful for discussions and comments regarding my thesis from Dr. Michael Schäfer.
Furthermore, many thanks go to the cloud physics group at IMK-TRO and to all

past and present colleagues who made the work joyful. I thank my office mates for the nice atmosphere, in particular Isabelle Reichardt for many discussion about work and future. Many thanks go to Dr. Christian Barthlott for always being approachable regarding the COSMO model and computer challenges, as well as for proof-reading a chapter of this thesis. Furthermore, I thank Dr. Marco Paukert for his very timely answers to all my model setup questions. A big share of my gratitude goes to Eva Hubel, Linda Schneider, and Constanze Wellmann for reading through parts of this thesis. I thank Thea Schiebel and Dr. Romy Ullrich for always having time and a place for me when I was visting AAF.
A special thank you goes to all colleagues on the 13th floor and at Campus North for the friendly atmosphere and to Dr. Gregor Pante and Dr. Florian Pantillon for forming the heart of the Tuesday coffee round. I thank Doris Stenschke, Friederike Schönbein, and Gerhard Brückel for their support in administrative and IT questions. Without them, focus on research would have been difficult. For keeping up the productivity, active lunch breaks were very helpful: Thanks for fun and action Hilke Lentink, Lisa-Ann Quandt, Linda Schneider, Dr. Florian Pantillon, Dr. Christian Barthlott, and many more.
I would like to thank all my friends in Karlsruhe and elsewhere to be part of my life besides PhD. Thank you Eva for always having time for discussing all kind of topics. Finally, I thank my parents Birgit Weixler and Thomas Weixler and my sisters Susanne Striegel and Franziska Weixler for supporting me, believing in me during all the years, and beeing merciful if I did not have time to answer phone calls. My special thank goes to my husband Axel, who always believes in me and helps me to see the positive things in life every day.

Wissenschaftliche Berichte des Instituts für Meteorologie und Klimaforschung des Karlsruher Instituts für Technologie (0179-5619)

Bisher erschienen:

Nr. 1: *Fiedler, F. / Prenosil, T.*
Das MESOKLIP-Experiment. (Mesoskaliges Klimaprogramm im Oberrheintal). August 1980

Nr. 2: *Tangermann-Dlugi, G.*
Numerische Simulationen atmosphärischer Grenzschichtströmungen über langgestreckten mesoskaligen Hügelketten bei neutraler thermischer Schichtung. August 1982

Nr. 3: *Witte, N.*
Ein numerisches Modell des Wärmehaushalts fließender Gewässer unter Berücksichtigung thermischer Eingriffe. Dezember 1982

Nr. 4: *Fiedler, F. / Höschele, K. (Hrsg.)*
Prof. Dr. Max Diem zum 70. Geburtstag. Februar 1983 (vergriffen)

Nr. 5: *Adrian, G.*
Ein Initialisierungsverfahren für numerische mesoskalige Strömungs modelle. Juli 1985

Nr. 6: *Dorwarth, G.*
Numerische Berechnung des Druckiderstandes typischer Geländeformen. Januar 1986

Nr. 7: *Vogel, B.; Adrian, G. / Fiedler, F.*
MESOKLIP-Analysen der meteorologischen Beobachtungen von mesoskaligen Phänomenen im Oberrheingraben. November 1987

Nr. 8: *Hugelmann, C.-P.*
Differenzenverfahren zur Behandlung der Advektion. Februar 1988

Nr. 9: *Hafner, T.*
Experimentelle Untersuchung zum Druckwiderstand der Alpen. April 1988

Nr. 10: *Corsmeier, U.*
Analyse turbulenter Bewegungsvorgänge in der maritimen atmosphärischen Grenzschicht. Mai 1988

Nr. 11: *Walk, O. / Wieringa, J.(eds)*
Tsumeb Studies of the Tropical Boundary-Layer Climate. Juli 1988

Nr. 12: *Degrazia, G. A.*
Anwendung von Ähnlichkeitsverfahren auf die turbulente Diffusion in der konvektiven und stabilen Grenzschicht. Januar 1989

Nr. 13: *Schädler, G.*
Numerische Simulationen zur Wechselwirkung zwischen Landoberflächen und atmophärischer Grenzschicht. November 1990

Nr. 14: *Heldt, K.*
Untersuchungen zur Überströmung eines mikroskaligen Hindernisses in der Atmosphäre. Juli 1991

Nr. 15: *Vogel, H.*
Verteilungen reaktiver Luftbeimengungen im Lee einer Stadt – Numerische Untersuchungen der relevanten Prozesse. Juli 1991

Nr. 16: *Höschele, K.(ed.)*
Planning Applications of Urban and Building Climatology – Proceedings of the IFHP / CIB-Symposium Berlin, October 14-15, 1991. März 1992

Nr. 17: *Frank, H. P.*
Grenzschichtstruktur in Fronten. März 1992

Nr. 18: *Müller, A.*
Parallelisierung numerischer Verfahren zur Beschreibung von Ausbreitungs- und chemischen Umwandlungsprozessen in der atmosphärischen Grenzschicht. Februar 1996

Nr. 19: *Lenz, C.-J.*
Energieumsetzungen an der Erdoberfläche in gegliedertem Gelände. Juni 1996

Nr. 20: *Schwartz, A.*
Numerische Simulationen zur Massenbilanz chemisch reaktiver Substanzen im mesoskaligen Bereich. November 1996

Nr. 21: *Beheng, K. D.*
Professor Dr. Franz Fiedler zum 60. Geburtstag. Januar 1998

Nr. 22: *Niemann, V.*
Numerische Simulation turbulenter Scherströmungen mit einem Kaskadenmodell. April 1998

Nr. 23: *Koßmann, M.*
Einfluß orographisch induzierter Transportprozesse auf die Struktur der atmosphärischen Grenzschicht und die Verteilung von Spurengasen. April 1998

Nr. 24: *Baldauf, M.*
Die effektive Rauhigkeit über komplexem Gelände – Ein Störungstheoretischer Ansatz. Juni 1998

Nr. 25: *Noppel, H.*
Untersuchung des vertikalen Wärmetransports durch die Hangwindzirkulation auf regionaler Skala. Dezember 1999

Nr. 26: *Kuntze, K.*
Vertikaler Austausch und chemische Umwandlung von Spurenstoffen über topographisch gegliedertem Gelände. Oktober 2001

Nr. 27: *Wilms-Grabe, W.*
Vierdimensionale Datenassimilation als Methode zur Kopplung zweier verschiedenskaliger meteorologischer Modellsysteme. Oktober 2001

Nr. 28: *Grabe, F.*
Simulation der Wechselwirkung zwischen Atmosphäre, Vegetation und Erdoberfläche bei Verwendung unterschiedlicher Parametrisierungsansätze. Januar 2002

Nr. 29: *Riemer, N.*
Numerische Simulationen zur Wirkung des Aerosols auf die troposphärische Chemie und die Sichtweite. Mai 2002

Nr. 30: *Braun, F. J.*
Mesoskalige Modellierung der Bodenhydrologie.
Dezember 2002

Nr. 31: *Kunz, M.*
Simulation von Starkniederschlägen mit langer Andauer über Mittelgebirgen. März 2003

Nr. 32: *Bäumer, D.*
Transport und chemische Umwandlung von Luftschadstoffen im Nahbereich von Autobahnen – numerische Simulationen.
Juni 2003

Nr. 33: *Barthlott, C.*
Kohärente Wirbelstrukturen in der atmosphärischen Grenzschicht. Juni 2003

Nr. 34: *Wieser, A.*
Messung turbulenter Spurengasflüsse vom Flugzeug aus.
Januar 2005

Nr. 35: *Blahak, U.*
Analyse des Extinktionseffektes bei Niederschlagsmessungen mit einem C-Band Radar anhand von Simulation und Messung.
Februar 2005

Nr. 36: *Bertram, I.*
Bestimmung der Wasser- und Eismasse hochreichender konvektiver Wolken anhand von Radardaten, Modellergebnissen und konzeptioneller Betrachtungen. Mai 2005

Nr. 37: *Schmoeckel, J.*
Orographischer Einfluss auf die Strömung abgeleitet aus Sturmschäden im Schwarzwald während des Orkans „Lothar". Mai 2006

Nr. 38: *Schmitt, C.*
Interannual Variability in Antarctic Sea Ice Motion: Interannuelle Variabilität antarktischer Meereis-Drift.
Mai 2006

Nr. 39: *Hasel, M.*
Strukturmerkmale und Modelldarstellung der Konvektion über Mittelgebirgen. Juli 2006

Ab Band 40 erscheinen die Wissenschaftlichen Berichte des Instituts für Meteorologie und Klimaforschung bei KIT Scientific Publishing (ISSN 0179-5619). Die Bände sind unter www.ksp.kit.edu als PDF frei verfügbar oder als Druckausgabe bestellbar.

Nr. 40: *Lux, R.*
Modellsimulationen zur Strömungsverstärkung von orographischen Grundstrukturen bei Sturmsituationen
ISBN 978-3-86644-140-8

Nr. 41: *Straub, W.*
Der Einfluss von Gebirgswellen auf die Initiierung und Entwicklung konvektiver Wolken
ISBN 978-3-86644-226-9

Nr. 42: *Meißner, C.*
High-resolution sensitivity studies with the regional climate model COSMO-CLM
ISBN 978-3-86644-228-3

Nr. 43: *Höpfner, M.*
Charakterisierung polarer stratosphärischer Wolken mittels hochauflösender Infrarotspektroskopie
ISBN 978-3-86644-294-8

Nr. 44: *Rings, J.*
Monitoring the water content evolution of dikes
ISBN 978-3-86644-321-1

Nr. 45: *Riemer, M.*
Außertropische Umwandlung tropischer Wirbelstürme: Einfluss auf das Strömungsmuster in den mittleren Breiten
ISBN 978-3-86644-766-0

Nr. 46: *Anwender, D.*
Extratropical Transition in the Ensemble Prediction System of the ECMWF: Case Studies and Experiments
ISBN 978-3-86644-767-7

Nr. 47: *Rinke, R.*
Parametrisierung des Auswaschens von Aerosolpartikeln durch Niederschlag
ISBN 978-3-86644-768-4

Nr. 48: *Stanelle, T.*
Wechselwirkungen von Mineralstaubpartikeln mit thermo-dynamischen und dynamischen Prozessen in der Atmosphäre über Westafrika
ISBN 978-3-86644-769-1

Nr. 49: *Peters, T.*
Ableitung einer Beziehung zwischen der Radarreflektivität, der Niederschlagsrate und weiteren aus Radardaten abgeleiteten Parametern unter Verwendung von Methoden der multivariaten Statistik
ISBN 978-3-86644-323-5

Nr. 50: *Khodayar Pardo, S.*
High-resolution analysis of the initiation of deep convection forced by boundary-layer processes
ISBN 978-3-86644-770-7

Nr. 51: *Träumner, K.*
Einmischprozesse am Oberrand der konvektiven atmosphärischen Grenzschicht
ISBN 978-3-86644-771-4

Nr. 52: *Schwendike, J.*
Convection in an African Easterly Wave over West Africa and the Eastern Atlantic : A Model Case Study of Hurricane Helene (2006) and its Interaction with the Saharan Air Layer
ISBN 978-3-86644-772-1

Nr. 53: *Lundgren, K.*
Direct Radiative Effects of Sea Salt on the Regional Scale
ISBN 978-3-86644-773-8

Nr. 54: *Sasse, R.*
Analyse des regionalen atmosphärischen Wasserhaushalts unter Verwendung von COSMO-Simulationen und GPS-Beobachtungen
ISBN 978-3-86644-774-5

Nr. 55: *Grenzhäuser, J.*
Entwicklung neuartiger Mess- und Auswertungs-strategien für ein scannendes Wolkenradar und deren Anwendungsbereiche
ISBN 978-3-86644-775-2

Die Bände sind unter www.ksp.kit.edu als PDF frei verfügbar oder als Druckausgabe bestellbar.

Nr. 56: *Grams, C.*
Quantification of the downstream impact of extratropical transition for Typhoon Jangmi and other case studies
ISBN 978-3-86644-776-9

Nr. 57: *Keller, J.*
Diagnosing the Downstream Impact of Extratropical Transition Using Multimodel Operational Ensemble Prediction Systems
ISBN 978-3-86644-984-8

Nr. 58: *Mohr, S.*
Änderung des Gewitter- und Hagelpotentials im Klimawandel
ISBN 978-3-86644-994-7

Nr. 59: *Puskeiler, M.*
Radarbasierte Analyse der Hagelgefährdung in Deutschland
ISBN 978-3-7315-0028-5

Nr. 60: *Zeng, Y.*
Efficient Radar Forward Operator for Operational Data Assimilation within the COSMO-model
ISBN 978-3-7315-0128-2

Nr. 61: *Bangert, M. J.*
Interaction of Aerosol, Clouds, and Radiation on the Regional Scale
ISBN 978-3-7315-0123-7

Nr. 62: *Jerger, D.*
Radar Forward Operator for Verification of Cloud Resolving Simulations within the COSMO Model
ISBN 978-3-7315-0172-5

Nr. 63: *Maurer, V.*
Vorhersagbarkeit konvektiver Niederschläge : Hochauflösende Ensemblesimulationen für Westafrika
ISBN 978-3-7315-0189-3

Nr. 64: *Stawiarski, C.*
Optimizing Dual-Doppler Lidar Measurements of Surface Layer Coherent Structures with Large-Eddy Simulations
ISBN 978-3-7315-0197-8

Nr. 65: *Mahlke, H.*
Mechanismen der Auslösung hochreichender Konvektion im südwestdeutschen Mittelgebirgsraum
ISBN 978-3-7315-0203-6

Nr. 66: *Fosser, G.*
Precipitation statistics from regional climate model at resolutions relevant for soil erosion
ISBN 978-3-7315-0227-2

Nr. 67: *Adler, B.*
Boundary-Layer Processes Producing Mesoscale Water-Vapour Variability over a Mountainous Island
ISBN 978-3-7315-0247-0

Nr. 68: *Kraut, I.*
Separating the Aerosol Effect in Case of a „Medicane"
ISBN 978-3-7315-0405-4

Nr. 69: *Breil, M.*
Einfluss der Boden-Vegetation-Atmosphären Wechselwirkungen auf die dekadische Vorhersagbarkeit des Westafrikanischen Monsuns
ISBN 978-3-7315-0420-7

Nr. 70: *Lott, F. F.*
Wind Systems in the Dead Sea and Footprints in Seismic Records
ISBN 978-3-7315-0596-9

Nr. 71: *Rieger, D.*
Der Einfluss von natürlichem Aerosol auf Wolken über Mitteleuropa
ISBN 978-3-7315-0672-0

Nr. 72: *Loewe, K.*
Arctic mixed-phase clouds. Macro- and microphysical insights with a numerical model
ISBN 978-3-7315-0686-7

Die Bände sind unter www.ksp.kit.edu als PDF frei verfügbar oder als Druckausgabe bestellbar.